A Brief History of the Earth's Climate

Praise for
A Brief History of the Earth's Climate

I love it. Earle understands the big climate picture and paints it with exceptional clarity.

> — James Hansen, director, Climate Science,
> Awareness and Solutions, Columbia University Earth Institute

An informative, succinct, and fascinating read—Steven Earle offers a unique and detailed account of Earth's climate history. His innate story-telling ability, coupled with his remarkable talent for making complex scientific information accessible, makes this page-turner a must-read for anyone seeking to understand the Earth's climate system.

> — Andrew Weaver, professor, University of Victoria,
> lead author, Intergovernmental Panel on Climate Change,
> second, third, fourth, and fifth Assessment Reports,
> former chief editor, *Journal of Climate*

An engaging tour through the complex natural processes at play in writing the Earth's long history of natural climate change to our present climate emergency. This primer will give campaigners, policymakers, and concerned citizens a more thorough understanding of climate science and renewed conviction to go all in on applying the brakes, leaving fossil fuels behind, and embracing a cleaner, healthier, and more equitable future.

> — Tom Green, Senior Climate Policy Advisor, David Suzuki Foundation

People interested in climate change, which these days should be everyone, need a basic understanding of the science of why Earth's climate is the way it is, and why it sometimes changes. Earle's book makes that complicated story easy to grasp. It's a model for clear science writing, and it forcefully awakens readers to what's at stake and what needs to be done.

> — Richard Heinberg, Senior Fellow, Post Carbon Institute, author, *Power*

A clear, concise and engaging introduction to the global ecosystem processes that govern our climate. A fascinating read for anyone ready to go beyond the headlines to learn more about how climate has shaped our history, why current climate change poses an unprecedented threat to our future, and what we can do about it.

— Laura Lengnick, author, *Resilient Agriculture:*
Cultivating Food Systems for a Changing Climate

A BRIEF HISTORY

OF THE

EARTH'S CLIMATE

EVERYONE'S GUIDE *to*
The SCIENCE *of* CLIMATE CHANGE

STEVEN EARLE, PHD

new society
PUBLISHERS

Cover design by Diane McIntosh.
Cover photo: ©iStock
All figures, drawings, and photos by author unless otherwise noted.

Printed in Canada. First printing September, 2021.

Inquiries regarding requests to reprint all or part of *A Brief History of the Earth's Climate* should be addressed to New Society Publishers at the address below. To order directly from the publishers, please call toll-free (North America) 1-800-567-6772, or order online at www.newsociety.com

Any other inquiries can be directed by mail to

New Society Publishers
P.O. Box 189, Gabriola Island, BC V0R 1X0, Canada
(250) 247-9737

LIBRARY AND ARCHIVES CANADA CATALOGUING IN PUBLICATION

Title: A brief history of the Earth's climate : everyone's
guide to the science of climate change / Steven Earle, PhD.

Names: Earle, Steven, author.

Description: Includes bibliographical references and index.

Identifiers: Canadiana (print) 20210264411 | Canadiana (ebook) 20210264713 |
ISBN 9780865719590 (softcover) | ISBN 9781550927528 (PDF) |
ISBN 9781771423489 (EPUB)

Subjects: LCSH: Climatic changes. | LCSH: Climatic changes—
Effect of human beings on. | LCSH: Global warming.

Classification: LCC QC903 .E27 2021 | DDC 363.738/74—dc23

Funded by the Government of Canada   Financé par le gouvernement du Canada

New Society Publishers' mission is to publish books that contribute in fundamental ways to building an ecologically sustainable and just society, and to do so with the least possible impact on the environment, in a manner that models this vision.

Contents

Preface

Let them not say: we did not see it.
We saw.
Let them not say: we did not hear it.
We heard.
Let them not say: they did not taste it.
We ate, we trembled.
Let them not say: it was not spoken, not written.
We spoke, we witnessed with voices and hands.
Let them not say: they did nothing.
We did not-enough.

— From "Let Them Not Say," by Jane Hirshfield[1]

Climate change is not coming; it is here now. The indications are clear, all around the world, with new ones coming to light virtually every day. One would need to have lucrative business interests, strong political convictions, or an impressive degree of stubbornness to be comfortable in saying that there is no strong evidence for anthropogenic (human-caused) climate change.

Those who do deny anthropogenic climate change often use the argument that the climate has changed before, and in that they are absolutely correct. The Earth's climate has been changing one way or another for 4.6 billion years. We have a reasonably good idea of how it has changed, and when, and why. The main natural mechanisms are changes in the sun's output, evolutionary changes in the lifestyles of organisms, moving continents and colliding tectonic plates, volcanic eruptions, incoming comets and meteorites, and even the Earth's variable orbit around the sun. Most of these changes have been excruciatingly slow, but some have been fast—even faster than anthropogenic climate change. Some are in the past, but most are still operating, and some of those do affect our climate on a human time scale.

The premise behind this book is that in order to fully understand anthropogenic climate change, we need to understand the Earth's long history of natural climate change. With just a limited knowledge of how the sun changes, how ocean currents behave, how the Earth wobbles (and why that matters), or how volcanic eruptions affect the climate, we can readily see that none of these natural processes can account for any part of the observed 1°C rise in the Earth's average surface temperature over the past 60 years. It's all on us.

Knowing something about past natural climate change is crucial to understanding the processes that are contributing to anthropogenic climate change now, how the forcing mechanisms—such as increased greenhouse gas levels—nudge the climate to a warmer state, and how the feedback mechanisms—such as melting ice—amplify those forcings. That knowledge of past climate changes should also help us to determine how close we are to a tipping point that could send the climate into an altered state, one that we will not be able to control.

Does 1°C of warming matter? After all, nobody really cares if tomorrow is a degree warmer than today. But this isn't about just one day; it's about it being 1°C warmer every day (on average). In any year, some days might still be cooler than the long-term average, and others might be about average, but we can expect that most days will be warmer, and some will be much more than 1° warmer, and that can make a huge difference. We can also expect dry places to be dryer and wet places wetter, and storms to be more intense, and of course sea level to rise, because ice is melting.

So yes, 1°C of warming really does matter if your children are starving because your crops have been shriveled by drought for the fourth year in a row. It does if your only supply of fresh water has dried up. It does if your life savings have been wiped out by an out-of-control wildfire. It does if your entire community has been destroyed by a flooding river, or a landslide, or a super storm named for a Greek letter. It does to hundreds of millions of people whose cities, farms, homes, or workplaces are threatened by sea-level rise. It does to an organism that can no longer survive in its ecosystem and has nowhere else to go, or no means to get there. And it also matters—even for those not facing any of these risks—if they have some understanding of the kind of treacherous and unpredictable terrain we're on here, and how close we might be to the edge of a cliff that we can't see.

About 20 years ago, at a time when I had only a vague concept of the terrain that is climate change, I was given the opportunity to develop and teach a course on environmental geology. As any teacher knows, there is no better way to learn about something than to try to teach it, and I soon learned that while the Earth, with its earthquakes, volcanoes, landslides, and floods, is a dangerous place to begin with, and that although we have created a wide range of nasty environmental problems, climate change is far more significant and more dangerous than all of the other threats. In fact, unless we come to grips with climate change, every other environmental threat will become largely irrelevant.

One degree of warming matters to me; it matters enough to make me alter my lifestyle significantly, to march around the streets with signs and noisemakers, and to put the time into writing this book. It matters because I can see how 1° of warming has already changed our world; and it matters even more because I fear the unknown terrain that we will be venturing into if we don't all make some big adjustments to the way we live, and very soon.

Please, let them not say, in 20 years or 50 years, that although we knew, we did not do enough.

— Steven Earle
December 2020

Acknowledgments

I am grateful to have been able to work on this book on the unceded traditional territory of the Snuneymuxw First Nation, who have lived in this region for many thousands of years, and whose lifestyles did not result in significant changes to the land or to the precious Salish Sea surrounding it, and who did not do things that could lead to changes to the climate.

I am also grateful to Isaac, Tim, Heather, and Justine for your valuable feedback on various parts of the manuscript, and to thousands of students for letting me be your guide along the rocky and sometimes fog-shrouded paths of earth science, and for all of your great questions and insights.

Introduction

This book is primarily about the natural processes of climate change that have operated on Earth for the past 4.6 billion years. It is critical to understand these natural phenomena in order to fully understand the processes of anthropogenic (human-caused) climate change that are operating now. That insight into the distant past also shows us that the climate changes that we have witnessed over the past century are not a result of natural climate forcing; they are entirely caused by us.

The book is organized on the basis of the time scales of the various natural phenomena, but it starts with an overview of the mechanisms that control Earth's climate, both now and in the past (chapter 1) and ends with a summary of what steps we can all take to reduce our personal and collective climate impacts (chapter 11).

In chapter 2, we look at the evolution of the sun over billions of years and how the Earth and its organisms have managed to keep the climate within a range that is suitable for life, in spite of a 40% increase in solar intensity.

Chapter 3 is focused on the painfully slow processes of plate tectonics, including how—over hundreds of millions of years—continental drift can control how much of the sun's energy gets converted into the heat, how tectonic processes can change the ocean currents that influence the climate, and how the formation of mountain ranges can change the composition of the atmosphere and therefore the climate.

In chapter 4, we consider the climate-cooling and climate-warming effects of volcanic eruptions, and the time scales at which they operate—years to tens of millions of years.

Chapter 5 provides an overview of the variations in the Earth's orbital parameters (Milanković cycles), how they have regulated the glacial cycles of the past million years, and whether or not we are headed into another glacial period.

In chapter 6, we look at the climate effects of long- and short-term changes to ocean currents, including the Gulf Stream in the Atlantic (which changes over hundreds of years) and the El Niño variations in the Pacific (which change over years).

Chapter 7 is focused on short-term solar cycles related to sunspot numbers, how they lead to small changes in solar output, and whether or not those changes (on time scales of decades) have implications for our climate.

Chapter 8 includes an examination of the catastrophic climate effects of collisions with large extraterrestrial bodies—such as the one that killed off the dinosaurs at the end of the Cretaceous Period (mostly over a period of several days)—along with a discussion of the probability that a similar event may happen in our future.

The climate implications of the activities of our *Homo sapiens* ancestors are summarized in chapter 9. Readers might be surprised to learn that there is evidence of human control over the climate going back several thousand years.

Chapter 10 is focused on tipping points. It includes a discussion of how and why the Earth's climate has tipped from one state to another in the past, and how some of the significant effects of anthropogenic climate change could lead to a tipping point in the near future.

It is no exaggeration to call anthropogenic climate change the most serious problem that humans have ever faced. The human and economic costs will be astronomical even if we make major changes now, but they will be many, many times worse if we continue to delay. The problem is not beyond our grasp, but it will require a collaborative and focused effort. Understanding some of the underlying natural processes will make it easier to understand why we all need to make changes.

1

WHAT CONTROLS
THE EARTH'S CLIMATE?

*We are running the most dangerous experiment in history
right now, which is to see how much carbon dioxide the
atmosphere can handle before there is an environmental
catastrophe.*

— Elon Musk, on Twitter, December 31, 2016

THE GREENHOUSE EFFECT, to which carbon dioxide is the main
contributor, is one of the key drivers of climate change, both now
and in the distant geological past, but there are other important
drivers, including changes in the amount of solar energy received at
different places on Earth, changes in the reflectivity (albedo) of Earth's
surfaces, and changes in the amount of particulate matter in the
atmosphere. These driving mechanisms are known as climate forcings, meaning that they force or nudge the climate to either a cooler
or a warmer state.

On the other hand, the real workhorses of climate change are positive feedbacks, which are natural processes that amplify the climate
forcings. For example, sea ice that is covered with snow is highly reflective. Most of the sunlight that hits it bounces straight back into
space, with almost no warming effect here on Earth. If that sea ice
melts, leaving exposed open water, most of the sunlight is absorbed
and converted into heat, warming up the water and the air above it,
and leading to more melting.

This chapter includes a description of how climate forcings work
and how feedbacks amplify them. Because the book deals with processes that are very slow and, in many cases, have taken millions or

even billions of years to have significant effects, this chapter includes an overview of geological time and an explanation of how painfully slow geological processes can have huge implications.

The Greenhouse Effect

Our atmosphere (the air we breathe) is dominated by the gases nitrogen and oxygen (as N_2, which makes up about 78%, and as O_2, which makes up about 21%). There is also about 1% argon (as Ar), and there are numerous gases that are present at much smaller concentrations, such as carbon dioxide (CO_2), neon and helium, methane (CH_4), nitrous oxide (N_2O), ozone (O_3), and others at even smaller concentrations (figure 1.1). Water vapor is also a very important atmospheric gas. Its concentration is quite variable, from about 0.01% at very low temperatures to over 4% at 30°C. As shown on figure 1.1, we use "parts per million" (ppm) to describe concentrations of the less abundant gases. If a gas has a concentration of 1 ppm that means there is one molecule of that gas for every one million molecules of air.[1]

All gases with two or more atoms vibrate, just like the molecules of all solids and liquids.

Two-atom gas molecules, such as nitrogen (N_2) and oxygen (O_2), can vibrate only by stretching (figure 1.2), and these vibrations are relatively fast. Molecules with three or more atoms (carbon dioxide,

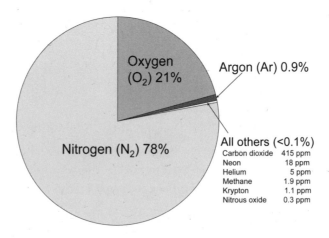

Figure 1.1. The composition of the lower atmosphere

for example) can vibrate both by stretching and by bending (figure 1.2). These are known as greenhouse gases (GHGs). The really important factor thing is that bending vibrations are slower than stretching vibrations, and the frequencies of bending vibrations fall within the range of frequencies of infrared radiation emitted from the warmed surfaces of the Earth.[2]

In order to understand the greenhouse effect, we must consider the different types of light. The light that comes to us from the sun is mostly in the visible part of the spectrum, although it extends into the near infrared and also a little way into the ultraviolet. The surfaces of the Earth (water, vegetation, ice, soil, and rock) are heated to varying

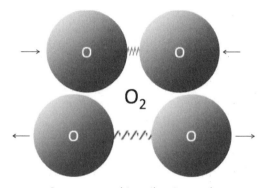

Oxygen: stretching vibrations only

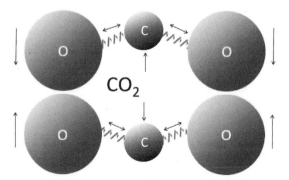

Carbon dioxide: stretching and bending vibrations

Figure 1.2. The vibrational modes of oxygen and carbon dioxide. The bending vibrations of carbon dioxide have frequencies that fall within the range of infrared radiation emitted by the Earth.

degrees by visible sunlight, and that makes them warm enough to emit light in the infrared (long-wave) part of the spectrum. That light is invisible to us, but you can see it in an infrared image, where the warmer the object, the brighter it will be.

Visible light from the sun vibrates at frequencies that don't match those of the common atmospheric gases (O_2 and N_2), or the GHGs, and so that light passes through our atmosphere without being absorbed (although it is reflected by clouds and particulate matter). But, as already noted, the frequency of infrared light radiated from the Earth's warmed surfaces does overlap with the bending-vibration frequencies of the GHGs, and when that light strikes those molecules, their vibrations become more vigorous. That warms the molecules, and that warms the air. In other words, the GHGs trap some of that infrared radiation from the warmed Earth. The higher the concentra-

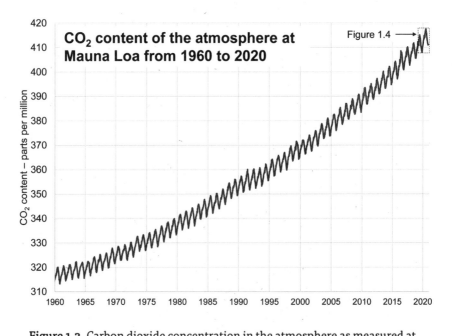

Figure 1.3. Carbon dioxide concentration in the atmosphere as measured at Mauna Loa, Hawaii, from Atmospheric CO_2 concentrations (ppm) derived from in situ air measurements at the Keeling Lab, Mauna Loa Observatory, Hawaii, operated by the Scripps Institute of Oceanography, U. of California, La Jolla, scrippsco2.ucsd.edu/data/atmospheric_co2/primary_mlo_co2 _record.html.

tions of GHGs, the more of that energy is trapped in the atmosphere, and the warmer it gets.

As most people are aware, carbon dioxide is the most significant of the GHGs. In late 2020, its concentration was about 415 parts per million (ppm), or 0.04% (figure 1.3), and it is currently increasing by about 2 ppm every year.

Why Is the CO_2 Curve So Squiggly?

As shown on figure 1.4, CO_2 levels at Mauna Loa peak in May of each year and then decrease to a minimum in September. This is because land plants grow vigorously from June through September, and that consumes a lot of atmospheric CO_2. Much of that is returned to the atmosphere as the organic matter breaks down in the fall and winter. But every May, a new peak is reached because of the massive amount of CO_2 that we emit by burning fossil fuels. CO_2 mixes readily in the atmosphere, especially in the east-west sense, so these results from Hawaii are generally representative of the northern hemisphere.

The pattern is the opposite in the southern hemisphere (the peak is in September each year), but the effect is not as strong because there is much less land south of the equator.

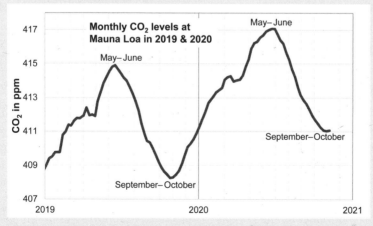

Figure 1.4. Carbon dioxide levels in 2019 and 2020. Source: See Figure 1.3.

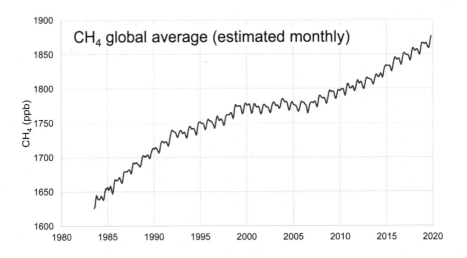

Figure 1.5. Globally averaged, monthly mean atmospheric methane abundance, from data by the Global Monitoring Division of the National Oceanic and Atmospheric Administration's Earth System Research Laboratory, E. Dlugokencky, NOAA/ESRL, esrl.noaa.gov/gmd/ccgg/trends_ch4.

The next most important GHG is methane, which is currently present at about 1,870 parts per billion (ppb), or 1.9 ppm, and is increasing every year by about 8 ppb (figure 1.5). Such a low concentration may seem to be insignificant compared with carbon dioxide, but methane is much more effective at absorbing infrared radiation than carbon dioxide, and that small amount accounts for about one-third of anthropogenic warming. Other important GHGs are nitrous oxide (N_2O), ozone (O_3), and the chlorofluorocarbons (CFCs).[3]

Water is also an effective GHG, but its climate implications are mixed because higher water levels correlate with greater cloud cover, and, as everybody knows, clouds are pretty good at blocking the sun. Most of the sunlight that hits the upper surfaces of the clouds is reflected back into space. The amount of water that the atmosphere can hold is proportional to the temperature, so warming does make the GHG effect of water more significant.

As we'll see in later chapters, GHG levels have varied widely in the past as a result of natural processes, including biological processes, volcanism, and weathering of rocks, and those variations have played a critical role in past climates. And, of course, GHG levels are changing significantly now because of human activities—mostly our use of fossil fuels and our dairy- and beef-rich diets.

Insolation

The strength of sunlight shining on the various surfaces of the Earth is known as insolation, and there have been significant changes in insolation over Earth's history. Chapter 2 includes a description of how the intensity of the sun has slowly changed over geological time, and how the Earth's systems have coped with that; chapter 7 includes a discussion of the climate implications of sunspot cycles that vary over periods from years to decades.

As we'll see in chapter 5, insolation changes are not just about how much energy is emitted by the sun. The amount of solar energy received on different parts of the Earth at different times of the year is affected by subtle changes in the shape of the Earth's orbit around the sun and the tilt of the Earth on its axis. This has significant climate implications and can lead to climate changes that are big enough to drive cycles of glaciation. That's because if there is less insolation in areas where glaciers are best able to form—around 60° north or south of the equator—there will be a tendency for glaciers to grow.

Albedo

The various surfaces of the Earth reflect light to differing degrees, and that property is known as albedo. In general, the darker a surface appears, the more light energy it absorbs, and that energy is converted into heat. Anyone who has walked barefoot on dark pavement in the hot sun knows this. As summarized on figure 1.6, ice and snow (especially fresh snow) and clouds have the highest albedo: between 70% and 90% of the light that hits these surfaces is reflected back into space and does virtually nothing to warm the Earth. Most rock and sand surfaces have albedos in the order of 30%, while forests are around 10 to 15%, and water is between 3% and 10%, but generally closer to 3% if the sun is overhead. Most of the light that hits these types of surfaces is absorbed and contributes to heating them. As we've just seen, those heated surfaces then emit infrared radiation, and that radiation interacts with GHGs to warm the atmosphere.

In the context of climate change, albedo matters only if it changes. There are lots of natural ways in which albedo can change; an obvious one is melting of snow or ice, which leads to lower albedo (because the exposed surface is darker) and a greater warming potential. Another is the loss of vegetation, which leads to a higher albedo in most cases (because bare ground is more reflective than vegetation) and, therefore, to cooling. (Yes, of course there are other factors involved here because healthy forests consume CO_2, and that role is more significant to the Earth's climate than their albedo. But from an albedo perspective alone, the loss of a forest leads to cooling.) As we'll see in chapter 3, continental drift can change the Earth's overall albedo if there is a net movement of continents into or out of tropical regions. That's because the climate implications of albedo differences are much more pronounced in the tropics than they are at higher latitudes and continents are more reflective than oceans. In other words, you get much more albedo bang for the buck near to the equator, where the insolation is most intense. Of course, such changes are painfully slow because continental drift takes place at rates of centimeters per year.

Many different human activities lead to albedo changes. Some examples include constructing paved roads and parking lots and build-

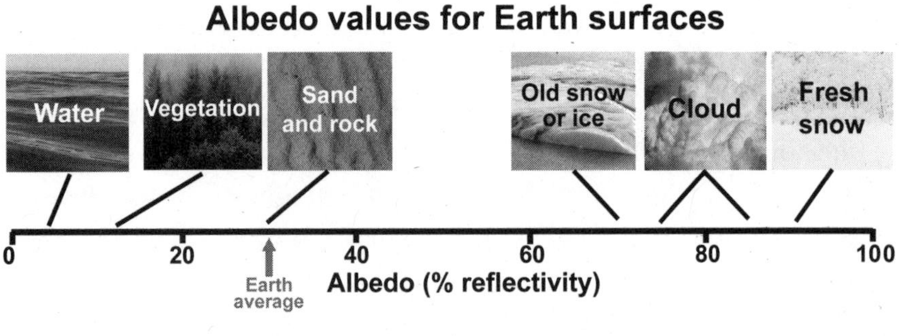

Albedo values for Earth surfaces

Figure 1.6. Typical albedos of some of Earth's surfaces

ings, cutting forests, growing crops, and producing smoke that coats snow and ice with particles of soot, thus diminishing their reflectivity.

Particulate Matter

Every year we pump millions of tonnes of particulate matter into the atmosphere, mostly as smoke from industrial operations and from motor vehicles.[4] This has a cooling effect because it blocks incoming sunlight. But it also has a warming effect because particles accumulating on ice and snow decrease the albedo.

There are also many different natural atmospheric particulates, including dust from windstorms, smoke from natural fires, and ash and sulphate aerosols from volcanic eruptions.

Feedbacks

A climate feedback is any process that can either amplify or dampen a climate forcing effect. A simple example is melting snow. When the temperature warms and enough snow melts to expose whatever is underneath it (e.g., bare ground or vegetation), the albedo at that location is decreased. As a result, more light can be absorbed and so the local area warms up more, and so more melting takes place and more light is absorbed, and so on. That's an example of a positive feedback. It will keep working in that way until there's no more snow to melt in that area.

When the CO_2 level in the atmosphere increases, plants grow better because they thrive on higher CO_2 levels, and so they consume more CO_2. That lowers the CO_2 level a little and so dampens the original effect. That's a negative feedback. On the other hand, if the CO_2 level continues to increase and the climate warms to the point where the existing vegetation communities can't thrive, they will consume less CO_2 and that will be a positive feedback (a stronger increase in atmospheric CO_2 levels).

Some of the important climate feedbacks are listed in table 1.1.

Most of these feedbacks work just as well in reverse during a period of climate cooling. For example, as the climate cools, more snow

Table 1.1. Important climate feedback mechanisms

Feedback	Mechanism (as climate warming takes place)	Pos/neg
Sea ice (or lake ice)	Sea ice melts to reveal open water. The albedo decreases, more solar energy is absorbed and so there is more melting.	Positive
Snow and glacial ice	Snow and ice melt to reveal bare ground or vegetation, the albedo decreases, more solar energy is absorbed, and so there is more melting.	Positive
Water vapor	Warm air can hold more water vapor, and that leads to more warming because water vapor is a GHG, although the effect is complicated by the cloudiness factor.	Positive
Carbon dioxide solubility	The capacity of the oceans to absorb CO_2 decreases with increasing temperature, and so, as ocean water warms, more of the huge ocean reservoir of CO_2 is released into the atmosphere, producing more warming.	Positive
Methane and CO_2 in permafrost	Warming leads to melting of permafrost, releasing stored methane and CO_2 into the atmosphere, and so more warming.	Positive
Vegetation growth (CO_2)	The higher CO_2 level that led to warming enhances plant growth, which consumes more CO_2, thus moderating the CO_2 increase.	Negative
Vegetation growth (albedo)	Enhanced vegetation growth makes a surface darker, so more solar energy is absorbed, leading to more warming.	Positive
Vegetation distress	Vegetation may become distressed by warming, so less CO_2 is consumed and there is more warming. (Where cooling causes vegetation distress, the feedback may be negative, as less CO_2 is consumed.)	Positive
Wildfire	Warming and regional drought increase the potential for wildfires, which result in CO_2 and particulate emissions and reduced CO_2 consumption until the forest starts to regrow.	Positive

(and perhaps glacial ice) will accumulate in some regions, increasing the albedo and leading to more cooling. Or, with cooling, more carbon dioxide gets dissolved in the oceans, and so the greenhouse effect is reduced, and cooling is enhanced.

The alarming thing about feedbacks is that almost all of them are positive, and so there is a strong tendency for a little bit of warming to be amplified into a lot of warming, and vice versa with cooling. In fact, if that wasn't the case, it's likely that many of the dramatic climate changes that have occurred would never have happened. For example, we might not have had multiple glaciations over the past million years, or we might have had nothing but glaciation for the past million years—and, therefore, might still be in the middle of a glacial period!

It is even more alarming that there is a potential for positive feedbacks to get out of control, and, as described in chapter 10, that can lead the climate over a tipping point and into a regime that is nothing like what we are used to, and from which there is no return on a human time scale. That is a place that we do not want to go!

Geological Time

There is no disputing that the Earth is old; the bigger problem is making sense of how old it is. Four thousand five hundred and seventy million years (or 4,570,000,000 years) is such a long time, and so much longer than a human's life—or even the span of all human lives—that none of us has a hope of really understanding what it means.

The geological time scale is a mechanism for visualizing Earth's history and for placing past events into a universal framework. The version shown on figure 1.7 provides some context for important events related to life on Earth, such as the first fish, the first land animals, the beginning and end of the dinosaurs, and the first members of the genus *Homo*.

One way to wrap your mind around geological time is to put it into the perspective of a single year, since we all know how long it is from one birthday to the next. If all of the Earth's 4,570,000,000 years were to be compressed into one year, each hour of that year would be

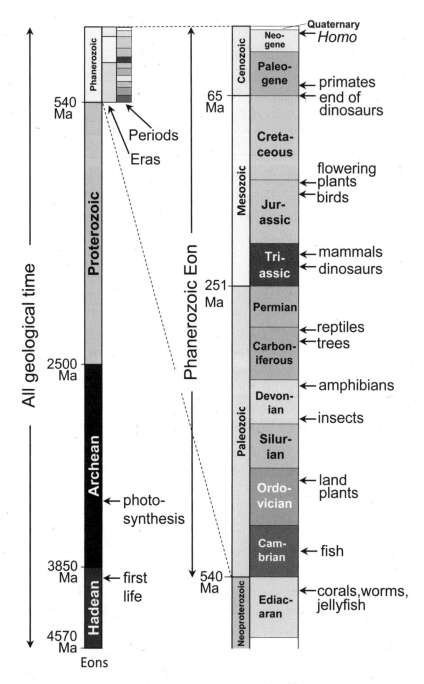

Figure 1.7. The geological time scale, and some important events in the history of life on Earth. Based on the International Commission on Stratigraphy (International Union of Geological Sciences), Cohen, K., et al., 2013, "The ICS International Chronostratigraphic Chart," *Episodes*, V. 36, pp. 199–204. stratigraphy.org/ICSchart/ChronostratChart2020-01.pdf.

equivalent to approximately 500,000 years of the Earth's history, and each day equivalent to 12.5 million years.

Using this analogy, we can say that the Earth formed on January 1. Life evolved in mid-February (around 4,000 million years ago), but there would have been nothing that was visible without a microscope until the ancestors of worms, jellyfish, and corals evolved on about November 13. Plants moved onto land around November 24 and amphibians on December 3. Reptiles evolved from amphibians during the first week of December. Dinosaurs and early mammals had evolved from reptiles by December 13, but the dinosaurs—which survived for 160 million years—were gone by Boxing Day (December 26). Primates evolved a day or so later (December 27), and humans from Asia first stepped foot into the western hemisphere at about two minutes before midnight on New Year's Eve.

Time is abundant, and that's a good thing because many of the processes that we're interested in are exceedingly slow. We often talk about slow processes happening "at glacial speed," but in geological terms, glaciers—which move at rates of meters to tens of meters per year—are really fast! Tectonic plates move at a few centimeters per year, and the sediments that turn into sedimentary rocks typically accumulate at less than 1 mm per year. Crystal growth is typically much slower still, in the order of millimeters per million years.

To put this into perspective, the plates that include the continents of Europe and North America are currently separating by about 2 cm/year, or about the length of the words "extremely slowly" on this page. The result of this process is the Atlantic Ocean, which is some 4,500 km wide, although it has taken about 200 million years to get that big.

I'm getting tired of typing "million years" so we're going to start using a shortcut to express geological time. Geologists use the abbreviations "Ma" (*mega annum*) to denote something that happened millions of years ago, and "Ga" (*giga annum*) for something that happened billions of years ago. So, the Earth originated 4,570 million years ago, or 4,570 Ma (which means the same as 4.57 Ga). Note that we don't have to say "4,570 Ma ago" because "ago" is implied with this notation. It's a bit like using time notation, such as "I have meeting at 9:30 am." On the other hand, just as you wouldn't say "My meeting will last for 2:00 pm," you wouldn't say "Dinosaurs existed for 149 Ma." Instead, you'd have to say "My meeting lasts for 2 hours" and "Dinosaurs existed for 149 million years" (because they existed from about 215 Ma to 66 Ma).

Climate-change Denial Arguments

Skepticism that the climate is actually changing, or—if it is—that humans are responsible, is widespread, and those who deny the whole concept of anthropogenic climate change use several arguments to support their case.

Lists of such arguments have been compiled by several organizations.[5] Some of the arguments that are pertinent to the subject of this book are as follows:

- It's the sun.
- The climate has always changed.
- Carbon dioxide levels are too low to make a difference.
- Climate models aren't accurate or reliable.
- There isn't consensus amongst climate scientists.
- It's related to volcanic eruptions.
- It's because of Milanković cycles.
- A warmer climate might be a good thing.
- We're heading for another ice age; this will prevent that.

It's worth looking at just one of those here because it is relevant to this chapter: the one about carbon dioxide levels being too low to make a difference. Carbon dioxide makes up only 0.04% of the atmosphere (or 415 ppm), so it is quite reasonable to question how it can have such a significant effect on our climate. However, there are several lines of evidence that support the conclusion that it does play a pivotal role:

- We know that CO_2 molecules can absorb radiation in the infrared part of the spectrum that the Earth emits, and that this leads to warming.

- Satellite observations show that this absorption is happening because infrared radiation from the Earth is depleted at the specific wavelengths that CO_2 absorbs.
- Similar observations show that this CO_2 spectral depletion of the infrared radiation has been increasing for several decades.
- There is a close correlation between increased CO_2 levels and warming over the past century, and the CO_2 levels measured are sufficient to explain the amount of warming observed.
- The other natural and anthropogenic changes that have happened over the same period cannot explain the warming that has been observed.

Some of the warming associated with increased CO_2 levels is a result of feedbacks. Warming is leading to destruction of permafrost in many temperate regions and to the consequent release of methane

Figure 1.8. A poster stuck to a lamp pole on a street in Vancouver. The small print at the bottom, added by someone else, reads "…over thousands of years. Humans did it in a few decades." Photo by Isaac Earle, November 2020.

and more carbon dioxide. Warming has also led to loss of snow and ice on land and at sea, and some of the warming we've experienced can be attributed to the resulting lower albedos.

The lamp-pole poster shown on figure 1.8 sums up climate-change skepticism nicely because the Milanković cycles are not responsible for any of the warming over the past century. The Milanković effect is described in chapter 5, and, as you'll see there, Milanković forcing has been toward slow cooling for the past several thousand years. Some of the other arguments of climate-change skeptics in the list above will be considered in other chapters.

2

A SLOWLY WARMING SUN

The Sun

Have you ever seen
anything
in your life
more wonderful

than the way the sun,
every evening,
relaxed and easy,
floats toward the horizon

and into the clouds or the hills,
or the rumpled sea,
and is gone—
and how it slides again

out of the blackness,
every morning,
on the other side of the world,
like a red flower…

From "The Sun," by Mary Oliver[1]

THERE ARE THREE ESSENTIALS to our life on this planet: one is clean liquid water; another is fresh air; and the third is the constant and reliable stream of radiant energy we get from the sun.

Evolution of the Sun

From the perspective of a human life span and human experience, the sun is indeed constant and perfectly reliable. There are some minor cyclical fluctuations in its output over periods of decades,[2] but those are not detectable without instruments or laborious observations. However, when considered over a much longer time frame—billions of years—our sun is anything but constant; it is slowly getting hotter. This chapter is about that change and its implications for the Earth's climate.

Based on centuries of observations, astronomers have pieced together a life cycle for stars like the sun, as shown on figure 2.1. Our star started forming at more than 4.6 Ga (4.6 billion years ago) when a massive cloud of gas—mostly hydrogen but with a small fraction of heavier elements—slowly collapsed due to its own gravity and eventually became dense enough that the hydrogen atoms were forced so tightly together that they began to fuse into helium atoms (nuclear fusion), thus giving off heat, and therefore light, and marking the birth of a star.

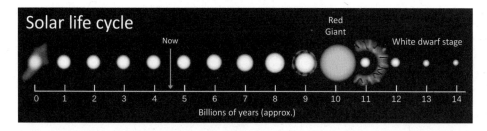

Figure 2.1. The life cycle of a main sequence star similar to our sun. Modified from Oliver Beatson, "Solar Life Cycle.svg," *Wikimedia*. Note that the depiction of the diameter of the sun is not to scale. During the red giant phase, the diameter of the sun is expected to be about 200 times larger than it is now, possibly big enough to engulf the Earth.

This process has continued for 4.57 billion years (to now) at a more-or-less-steady state, but as already noted, the sun has been gradually getting hotter and will continue to do so. This increase in intensity is a result of the ongoing conversion of hydrogen to helium in the sun's core. The growing proportion of helium results in an increase in the density of the solar core region, which causes the core to contract. The increased gravitational pressure forces the hydrogen atoms closer together, and that accelerates the rate of fusion and makes the sun hotter and brighter.[3]

The sun will continue to get hotter in this way for another 4 billion years or so, until its core is made up entirely of helium, at which point it will evolve into a red giant and will start to expand, first consuming Mercury and then Venus and quite likely even the Earth. At that time, the helium will start to fuse into carbon; almost half of the sun's mass

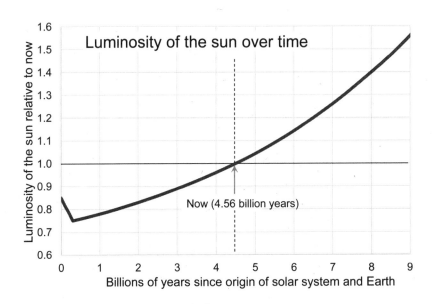

Figure 2.2. Variation in the sun's luminosity over 9 billion years. Based on information in Ribas, I., "Solar and Stellar Variability: Impact on Earth and Planets," *Proceedings of the International Astronomical Union Symposium, Solar and Stellar Variability: Impact on Earth and Planets*, V. 264: pp. 3–18, 2010.

will be lost into space via massive explosions, and what remains will eventually collapse into a white dwarf.

The change in solar luminosity[4] over 9 billion years of our sun's evolution is shown on figure 2.2. The amount of energy produced by the sun has increased by about 33% over its entire 4.6-billion-year history—so far. That's a huge amount from the perspective of Earth's climate, but it's over a very long time. The sun is going to get hotter still. Within another 4.4 billion years, it will be roughly twice as hot as it was originally.

But wait! Before anyone starts thinking that a warming sun is behind the current climate change, or a reason for us to be concerned for our distant future, we need to put this into perspective. The present rate of solar warming is about 8% every billion years. That's 0.008% every million years or 0.0000008% every century. That is a very, very small amount of warming on a human time scale. Over the past century, it has not been enough to noticeably warm the climate—not even close. For example, the increase in luminosity due to long-term solar evolution from 1920 to 2020 was only enough to increase the Earth's surface temperature by about 0.0000016°C. During that time, the surface temperature[5] actually increased by about 1°C, so it's clear that solar evolution is not the cause.

The long-term warming of the sun has no connection to the short-term sunspot cycle. That topic is addressed in chapter 7.

Early Life and Atmospheric Change

The earliest evidence of life on Earth is found in rocks about 4 billion years old.[6] At that time, the sun was about 80% as bright as it is today. An Earth with today's atmosphere, and with an 80% sun, would have been completely frozen (no liquid water anywhere on the surface)! That might lead us to wonder how it could have been possible for any type of life to evolve, and this is known as "the faint young sun paradox." The Earth wasn't encased in ice 4 billion years ago because the Archean[7] atmosphere was very different from today's atmosphere. It was considerably thicker—similar in density to that of the present atmosphere of Venus—and, as shown on figure 2.3, it was rich in carbon dioxide. The carbon dioxide proportion might have been in the order of 10%, as compared with today's level of 0.04% (415 parts per million), and that was sufficient to keep the Earth's surface temperature warm even with a relatively faint sun.

Life on Earth may have evolved in a warm ocean, in a seafloor volcanic region close to scalding submarine hot springs, beneath an atmosphere rich in carbon dioxide and with no free oxygen at all (figure 2.4), or it may have evolved in a shallow pool that was subject to repeated wetting and drying.[8] In either case, free oxygen (such as the O_2 in our present atmosphere) would have been a deadly poison to the earliest microorganisms, just as it is today to some of the things that thrive in the dark and damp places in (and on!) our bodies and our homes, and in boggy environments around us. Many early lifeforms were methanogens—meaning that they produced methane—and so methane levels in the atmosphere increased as life started to thrive.

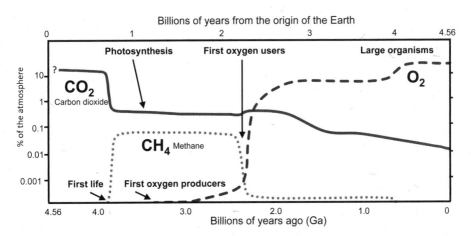

Figure 2.3. Evolution of the Earth's atmosphere over geological time. Based on information in Nisbet, E. and Fowler, C., "The Evolution of the Atmosphere in the Archaean and Early Proterozoic," *Chinese Sci Bull*, V. 56, pp. 4–13, 2011; and Large, R., et al., "Atmosphere Oxygen Cycling Through the Proterozoic and Phanerozoic," *Mineral Deposita*, V. 54, pp. 485–506, 2019.

This contributed to the warming because methane is a potent greenhouse gas.

Around 3.5 Ga, microorganisms developed the ability to use the sun as an energy source. The first of these were bacteria[9] (possibly cyanobacteria, aka blue-green algae), and they had a distinct advantage over other lifeforms of the day in that they didn't have to rely on energy from volcanic hot springs or from chemical sources: they could live virtually anywhere that the sun's light could reach. The photosynthetic process involves consumption of carbon dioxide and

Figure 2.4. Seafloor hydrothermal vent on the East Pacific Rise near Mexico. Photo by W. R. Normark and D. Foster of the U.S. Geological Survey (library photo.cr.usgs.gov) taken in 1981 at a depth of about 2,000 m on the East Pacific Rise, at 21° north (offshore from Baja California Sur). It shows a "black smoker" chimney. The hot water, which may be around 300°C, is coming from a seafloor vent, and it is black because sulphide minerals are precipitating from it due to rapid cooling in the cold sea water. There is no light at this depth, and all of the organisms living around the vent are ultimately dependent on the heat and chemicals in the water, which is hot because of nearby volcanic activity.

release of oxygen, but the proportion of free oxygen in the atmosphere remained very low for almost another billion years. That's because any oxygen produced by photosynthetic organisms was first used up in chemical reactions with abundant elements like iron or gases like methane and was also consumed through the decay of dead organic matter.

Free oxygen first started to accumulate in our atmosphere around 2.4 Ga, initially in very small amounts. We oxygen breathers might think that most of the organisms of the day would have rejoiced at the change, but of course they did not because these organisms weren't practiced in rejoicing, but much more importantly, they were still dependent on an oxygen-free environment. We call this transition the "oxygen crisis" because it led to extinction for many organisms. On the other hand, it really was the beginning of the beginning for organisms like us because the oxygen crisis appears to have pushed some existing organisms to evolve cells with a nucleus. Such organisms are eukaryotes, and they are our ancestors.

As more and more oxygen was produced, it continued to react with methane, and the methane level dropped.[10] Since methane is a powerful greenhouse gas, the climate cooled dramatically, triggering an extensive period of glaciation starting around 2.3 Ga, the Huronian glaciation.[11] That was followed by a very warm period, and then the climate levelled out a bit, and there is no evidence of glaciation on Earth for another 1.6 billion years. More on that in chapter 3.

So, to summarize, it was relatively warm during most of the Earth's first few billion years, despite a cooler sun. That's because greenhouse gas levels were much higher than they are now. But there's still a paradox here because life has evolved and flourished in liquid water (not frozen and not boiled off into space) for 4 billion years. How could the Earth have maintained a reasonable temperature throughout its history, a temperature moderate enough (notwithstanding some ups and downs, such as the Huronian glaciation) for some water to remain liquid? In other words, why has the Earth's atmosphere changed as the sun has warmed, in such a way that a "Goldilocks" climate has existed here for 4 billion years?

Gaia and the Earth's Climate

Although we don't fully understand why the Earth has been so habitable for so long, a key mechanism is the evolution of the atmosphere, and one of the drivers of that is photosynthesis. Various types of photosynthetic organisms—large and small, on land and in the oceans—have taken carbon dioxide from the atmosphere and released oxygen, converting the carbon to hydrocarbons and storing it in the rocks of the Earth's crust. A parallel process, with a similar outcome, is the conversion of carbon from carbon dioxide into carbonate minerals, which is what most shelled organisms do to make their shells. Over time, as the sun has slowly warmed, these two processes have reduced the greenhouse effect enough to keep the Earth from getting too hot.

The idea that life has controlled Earth's climate to its own benefit forms the basis of the Gaia theory, first proposed by James Lovelock in 1972[12] and expanded upon by Lovelock and Lynn Margulis in 1974.[13] According to Lovelock and Margulis, the Earth and the living organisms on it form a self-regulating system that ensures not only that conditions remain suitable for life to persist but also that they have been able to do so even as the system's main source of energy (the sun) has slowly changed in intensity. This self-regulation proceeds through various types of biological processes and climate feedbacks. (Climate feedbacks are discussed in chapter 1.)

Here's an example of how this can work. Imagine a vast population of bacteria living in the ocean in a warm climate. Most of them are the old-style bacteria, and a few are the new-improved photosynthetic bacteria that use the sun's energy to help them convert carbon dioxide to oxygen. Let's say that they thrive best in slightly cooler conditions, while the regular bacteria prefer it to be warm. As time passes (lots of it), the photosynthetic bacteria gradually consume enough carbon dioxide to cool the climate just a little bit. This is good for them and they thrive, while the regular bacteria shiver and complain (and die). In their increasing numbers, the photosynthetic bacteria cool the climate even more, so they thrive even better, and in time (lots of it), they start to dominate the bacteria population and continue to cool the climate. If they then become too successful and cool the climate too much for their own good, they will no longer flourish, and it will stop getting cooler.

Lovelock's Gaia theory was not well received by the scientific community in the early days. In fact, it went one worse than that: it was almost universally ignored. One issue was the use of the term "Gaia" (a Greek goddess) and Lovelock's musing that she was a living being. In 1979, he wrote: "But if Gaia does exist then we may find ourselves and all other living things to be parts and partners of a vast being who in her entirety has the power to maintain our planet as a fit and comfortable habitat for life."[14] This type of language makes most scientists cringe. Another serious concern was the implication that the organisms of the day had conceived a plan to make the climate amenable and set about doing it.

Some kind of "purpose" on the part of living organisms was not a component of Lovelock's theory at all, but the early writings on Gaia did not make that clear enough for many. A further concern was that Gaia wasn't a real scientific theory because it wasn't testable. Moreover, at that time many scientists who thought about the ancient climate just assumed that it was controlled purely by physical and chemical processes.

In 1983, Lovelock and Andrew Watson created a model called Daisyworld[15] to test the Gaia theory. The premise was an Earth-like planet populated only by daisies: white ones and black ones. Like ours, the planet was orbiting a warming star. The white daisies reflected solar energy and so had a net cooling effect on the planet, while the black ones absorbed solar energy and had a net warming effect. The bare surface of the planet had an albedo (reflectivity) that was midway between that of white and black daisies. On Daisyworld, the daisies can survive within the temperature range of 5° to 40°C, but they do best at 22.5°C. Using a numerical model, Watson and Lovelock showed that very early in the planet's history, while the star was still cool, Daisyworld was too cold for daisies to grow at all. As the planet's surface temperature eventually reached 5°C, they started to grow; black daisies did best because they had the effect of absorbing sunlight and warming their local environment, while white daisies did poorly because they cooled their local environment (figure 2.5). So black daisies ruled. As the star became progressively warmer, and there was less need to enhance that warmth, the white daisies started

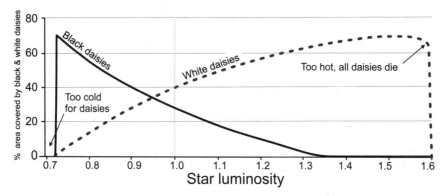

Figure 2.5. Black and white daisies as a proportion of surface area on Daisy-world as a function of luminosity (with respect to current luminosity of 1.0). Based on Watson, A., and Lovelock, J., "Biological Homeostasis of the Global Environment," 1983.

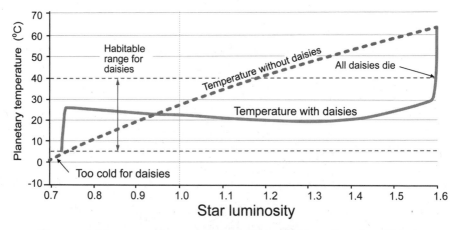

Figure 2.6. Mean annual temperature on Daisyworld as a function of luminosity. The dashed blue line shows how the temperature would naturally increase because of the increase in stellar luminosity. Based on Watson, A. and Lovelock, J., "Biological Homeostasis of the Global Environment," 1983.

to compete better with the black daisies, and with continued warming, the white daisies ruled. Eventually the star got so hot that even the white daisies couldn't reflect enough light away, and the temperature exceeded 40°C, resulting in the death of all daisies.

Figure 2.6 shows the temperature history of Daisyworld. As soon as the temperature reaches 5°C, the daisies start to grow, and black daisies quickly become dominant. Within a few million years, they have stabilized the temperature at close to the ideal of 22.5°C. From that point on, a combination of white and black daisies keeps the temperature near to that ideal, but when the star becomes too hot for even 100% white daisies to regulate the temperature, things go badly.

Although Daisyworld is a highly simplistic model for climate evolution, it demonstrates that this type of biological control over the climate is possible and also that it doesn't take purposeful or organized behavior to make it happen. Over the past few decades, the Gaia theory has gradually gained acceptance, and it is now firmly established as a viable explanation for evolution of the atmosphere and for Earth's 4.57-billion-year history of a (relatively) stable climate. Without mediation by life, there is essentially zero chance that the Earth's surface temperature would be almost the same now as it was at 4 Ga.

Figure 2.7. Rearguard Mountain (*left*), primarily made of limestone, just like much of the Canadian Rockies. A limestone outcrop in Belize (*right*), showing the shell components of the rock.

Storing Carbon

For the most part, biological control of the Earth's climate to compensate for a warming sun has been achieved—not by black and white daisies but by changes to the proportions of atmospheric gases, or more specifically by reductions in the concentrations of greenhouse gases. This means that most of the trillions of tonnes of carbon that used to be in the atmosphere as carbon dioxide and methane were slowly, methodically, and safely stored in the Earth's crust in rocks like limestone[16] (figure 2.7) and as hydrocarbon molecules in coal, oil, and natural gas.

Limestones have been created by a wide range of mostly marine organisms, including corals, bivalves, gastropods, cephalopods, sponges, arthropods, and algae. The organic matter that has become fossil fuels has been derived mostly from microorganisms, green algae, red algae, and later on both primitive and evolved land plants.

This storage of carbon in rocks has been taking place for a very long time. The oldest limestones are older than 3.5 Ga, and there are also carbon-rich black shales that date back to at least 3 Ga. Although most fossil-fuel deposits are in rocks younger than 250 Ma (250 million years), some are much older than that, even Precambrian (older than 540 Ma). That said, most of the really old rocks have been buried sufficiently deeply in the crust that any carbon they might have contained has been converted to graphite[17] at high temperatures.

The carbon in limestone and in fossil-fuel bearing rocks is stored "safely" because these rocks are mostly safe beneath the Earth's surface. Only a tiny proportion of such rock is naturally eroded each year, releasing some of the carbon to the atmosphere, while a similar amount is being stored away somewhere else, mostly on the seafloor. The problem is that humans are interfering with this natural process in a major way by digging up and burning the fossil fuels and using the limestone to make cement.

Every year, we use a massive amount of fossil fuels (including coal, oil, and natural gas), an amount equivalent to 80 billion barrels of oil. That's roughly the same as 115 million railway tanker cars, or an oil train long enough to go around the world 17 times! Every single day, that's equivalent to about 319,700 railway tank cars, enough to stretch from Toronto to Dallas or from Paris to Kiev. Another 2,000-kilometer-long train every day! Almost all of the carbon in that coal, oil, and gas ends up in the atmosphere as carbon dioxide: about 100 million tonnes every day. That very carbon was removed from the atmosphere tens and hundreds of millions of years ago, offsetting the impact of a warming sun, and it was stored in the rocks. By digging it up or pumping it out and putting it back into the atmosphere, as carbon dioxide, we are severely compromising the Earth's ability to regulate the temperature under a sun that is now much warmer than it was when that carbon was first stored away.

Future Solar Warming

What of the distant future? As the sun continues to warm over the next few billion years, the atmosphere must evolve toward a weaker greenhouse effect. If not, whatever is living on Earth at the time will find it difficult to survive. But this is not something we need to worry about because the sun's warming process is so slow. Over the next million years, the amount by which the sun is expected to warm represents a potential temperature change on Earth of about 0.016°C (if everything else is held constant). If humans haven't been voted off the planet by then, it's likely that this is something they'll be able to cope with.

But if we look ahead on a different time scale, things do get more difficult. Modeling future climates out to about 1.5 billion years shows that although the Earth should still support life at that time (because there should still be liquid water), the only place that anyone could survive would be on Antarctica.[18] But that's so far into the future that it's not even worth speculating what life on Earth might look like.

Sometime after that, the planet will become so hot that water will be boiled off into space, and then it will be truly uninhabitable for life as we know it.

3

SLIDING PLATES AND COLLIDING CONTINENTS

*The concept of continental drift first came to me as far
back as 1910, when considering the map of the world,
under the direct impression produced by the congruence
of the coast lines on either side of the Atlantic. At first,
I did not pay attention to the idea because I regarded
it as improbable. In the fall of 1911, I came quite
accidentally upon a synoptic report in which I learned for
the first time of palaeontological evidence for a former
land bridge between Brazil and Africa. As a result, I
undertook a cursory examination of relevant research
in the fields of geology and palaeontology, and this
provided immediately such weighty corroboration that
a conviction of the fundamental soundness of the idea
took root in my mind.*

— Alfred Wegener, 1929[1]

FROM 1915, WHEN THE German meteorologist Alfred Wegener first
advanced his concept of moving continents, through 1930, when
he died while conducting scientific work in the middle of Greenland,
and then until about 1965, most of the geoscience establishment paid
little attention to his proposition on continental drift. This indiffer-
ence wasn't entirely based on intransigence or xenophobia (although
both played a role); in fact, Wegener was so far ahead of his time that

nobody—including Wegener himself—understood the Earth well enough to be able to imagine how continents could move.

That gradually changed between 1915 and 1965 as we learned more and more about the processes taking place within the Earth's interior. Some of the important discoveries were as follows:

- There is a tremendous amount of heat stored deep within the Earth, and that additional heat is continually being produced by radioactive decay.
- The transfer of heat from the core to the mantle causes slow convection in the strong but plastic rock of the mantle.
- Convection is the driving force behind the motion of the plates that comprise the upper rigid layer (about 100 km thick) and of the continents that are parts of those plates.

It is now accepted that the plates move around on the surface at rates of a few centimeters per year, that new oceanic crust is formed in some places, while old oceanic crust is destroyed in others, and that processes taking place at plate boundaries can account for most earthquakes and volcanic eruptions as well as the construction of almost all mountain ranges and all ocean basins. These processes are understood to have been taking place for at least the past billion years, and possibly for most of geological time.

Plate tectonics is important to Earth's climate, and over geological time, plate-related processes have triggered some dramatic and highly consequential climate changes. Some of the reasons for these changes are as follows:

- The movement of continents can influence albedo-related[2] climate forcing because land is more reflective than ocean water and albedo (reflectance) differences have a greater effect at low latitudes than at high latitudes.
- Volcanism (which is commonly a product of plate processes) can lead to both short-term cooling and long-term warming (see chapter 4).
- Formation of mountain ranges, through plate tectonics, leads to cooling because mountains are more readily eroded than plains,

erosion contributes to weathering of rock, and weathering of rock consumes atmospheric CO_2 (more on this below).

- Reconfiguration of ocean basins leads to changes in ocean currents that can have significant climate implications.

Albedo and Continental Drift

As discussed in chapters 1 and 2, land surfaces are more reflective than open water, and some land surfaces are more reflective than others. Most snow- and ice-covered surfaces have albedos between 70% and 90%; most unvegetated surfaces are between 15% and 40% (lower if wet); while most vegetated surfaces are between 10% and 20%. Open water of oceans or lakes has a reflectivity close to 3% at high sun angles.

A critical factor in the context of albedo is latitude. Albedo makes a much bigger difference at low latitudes (equatorial regions), where solar intensity is high throughout the year, than it does at high latitudes, where solar intensity isn't very high—even in the summer—because the sun never gets very far above the horizon. At present, the continents (which make up 29% of the Earth's surface) are almost evenly distributed in a latitudinal sense, with approximately 33% in equatorial regions (between 30° north and 30° south), 38% in temperate regions (between 30° and 60° north and south), and 29% in polar regions (north of 60° north and south of 60° south).

At approximately 720 Ma, the situation was much different, as shown on figure 3.1. Most of the land area was part of the supercontinent Rodinia, with 50% of it in equatorial regions, 40% in temperate regions, and only 10% in polar regions. That much land in the sensitive equatorial zone had a cooling effect because of the higher albedo of the land (which was higher than it is now since there was no land vegetation at that time[3]) compared to the ocean and also because of the more intense rock weathering at low latitudes (more on that later in this chapter). The albedo implication of a supercontinent centered on the equator—more sunlight reflected back into space—is considered to be an important contributor to the first of the Cryogenian Period Snowball Earth glaciations.[4] Although we don't know exactly where the continents were in earlier times, there is no evidence of a clustering of continents near to the equator in the 2 billion years prior to the Cryogenian Period,[5] or since that time.

As noted in chapter 2, the Earth's first major glaciation was the Huronian, at around 2.3 Ga. The Earth was essentially free from glaciation for about 1,600 million years following that, but starting about 720 Ma, we entered the most intensive and the most extensive glacial period in our long history. The small temperature forcing caused by the albedo effect of an equatorial supercontinent (estimated at about 3°C of cooling) likely led to snow and ice accumulation at high elevations and higher latitudes. Lower temperatures also favored transfer of more of the atmosphere's CO_2 into the oceans. Those feedbacks soon drove more intense cooling, and before too long, the land was mostly glaciated. This is known as the Sturtian glaciation, which lasted about 60 million years. For much of that time, the Earth's mean annual temperature was about minus 40°C, and the entire ocean— even at the equator—was covered in more than 200 meters of ice. Because there was little liquid water anywhere on the Earth's surface,

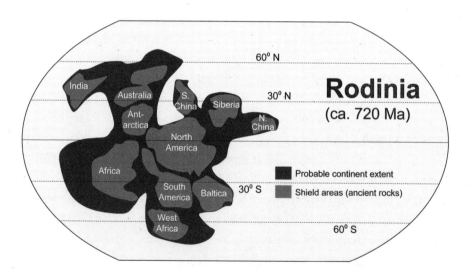

Figure 3.1. The likely latitudinal position of the supercontinent Rodinia at around 720 Ma. Based on Hoffman, P., et al., 2017, "Snowball Earth Climate Dynamics and Cryogenian Geology-geobiology," *Science Advances*, V. 3, pp. 1–43. You'll find lots of information at snowballearth.org.

the hydrological cycle was essentially shut down. Without much snow to feed terrestrial glaciers, they slowly lost ice by sublimation (direct evaporation), and some of the rocky surface was exposed.

Sixty million years is a very long winter (even by *Game of Thrones* standards), but considering that the bright icy surface reflected most of the incoming solar energy, it might have been longer still (perhaps even billions of years). Our saving grace is that the Earth's internal heat engine still motored on over that time, and volcanoes continued to erupt. Along with those eruptions came gases, including CO_2, and because there was no open ocean water and weathering rates were slowed by cold temperatures (more on that below), almost all of that volcanic CO_2 stayed in the atmosphere, gradually building a greenhouse effect strong enough to start melting the ice. It is likely that the CO_2 level had to reach about 13% (about 325 times the current level of 0.04%) in order to overcome the cooling effect of the bright surfaces.[6] As some of the terrestrial glaciers receded, positive feedbacks started working to enhance the warming, including the decrease in albedo caused by melting ice and the release of both carbon dioxide and methane from melting permafrost.

Eventually the sea ice started to melt, and it was probably mostly gone within several thousand years.[7] This relatively rapid transformation from reflective ice and snow to dark open water for the entire ocean, under an atmosphere with at least several percent CO_2, would have then contributed to an intense hothouse climate for at least another several thousand or tens of thousands of years.

Continental Collisions and Mountain Ranges

In plate tectonics, plates are either moving away from each other (diverging), moving toward each other (converging), or moving side by side (transforming). At most convergent plate boundaries, oceanic crust is being pushed down (subducted) beneath either continental crust or other oceanic crust. Oceanic crust is denser than continental crust because the seafloor is made of heavier rocks, and so while oceanic crust can be subducted, continental crust cannot. Getting continental crust to subduct would be a little like trying to push a large inflated toy down into the water of a lake or swimming pool.

Many oceanic crustal plates also include some continental crust (as islands or actual continents), and the two parts move along together. At about 100 Ma, the plate carrying the Indian continent started diverging from Antarctica and moved north toward Asia (figure 3.2). While that was happening, the continents were eroding and sediments and sedimentary rocks were accumulating on the ocean floor adjacent to both continents. When India reached Asia, sometime between 55 and 45 Ma, the continental part of the Indian plate was unable to subduct; instead, the rocks making up northern India and southern Asia, plus the sedimentary rocks in between, got crumpled, folded, faulted, and uplifted to start construction of what is

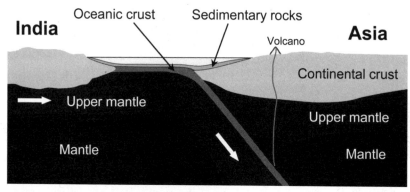

Figure 3.2. Representation of a subducting oceanic plate carrying the Indian continent north toward Asia at around 60 Ma.

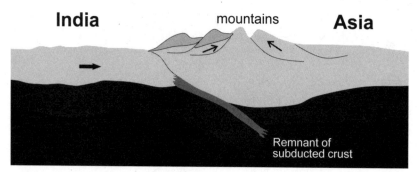

Figure 3.3. Representation of the India-Asia continental collision that led to the formation of the Himalayan mountains, starting around 50 Ma.

now—by a very wide margin—the Earth's highest and most extensive range of mountains (figure 3.3).[8] This uplift continued for tens of millions of years. In fact, the Indo-Australian Plate is still moving north, and still pushing the mountains up. So, if you've had a lifelong dream of climbing Mount Everest, you should get on it soon; the world's tallest mountain is getting higher by a few centimeters every year.

The Himalayans aren't the only significant mountains to have been formed in relatively recent geological times. Others include the Zagros and adjacent mountain ranges of Iran, Iraq, and Turkey, and the Alps of Europe, which were built mostly within the period of 65 to 40 Ma. All of these ranges can also be attributed to continental collisions.

Mountainous parts of the continents erode many times faster than plains (figure 3.4). The Himalayan range, which extends over 2,400

Figure 3.4. Rugged terrain in the Himalayan Annapurna region of Nepal, where rapid erosion is clearly evident from the thick accumulations of loose rocks on the lower slopes. Photo by Isaac Earle.

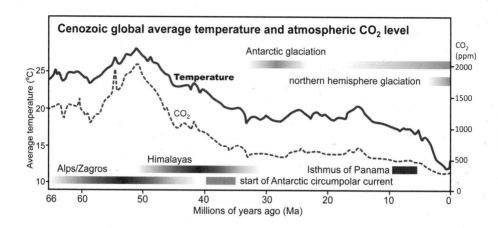

Figure 3.5. Global temperature (left scale) and atmospheric CO_2 level (right scale) for the past 66 million years, along with timing of mountain range formation, ocean current changes, and glaciations. Based on information from studies by James Hansen and others compiled by Root Routledge. See alpineanalytics.com/Climate/DeepTime.html.

kilometers from Myanmar to Pakistan and well north into southern China, is eroding faster than any similar sized area on the planet and has been doing so for close to 50 million years. One of the processes associated with that erosion is chemical weathering of the rocks. This takes many forms, but the one of interest to us here is the hydrolysis[9] of silicate minerals, such as feldspar, to form clay minerals—and the resulting consumption of atmospheric carbon dioxide.

The relationship between mountain building and global temperatures during the Cenozoic is illustrated on figure 3.5. Temperatures were consistently high through the Mesozoic (from 261 to 66 Ma), and that continued into the early part of the Cenozoic, but the climate started to cool around 50 Ma, and since then there has been a cumulative drop in global temperatures of about 14°C. This long-term decline closely follows the atmospheric CO_2 curve, and most of that change can be attributed to the enhanced weathering associated with mountain ranges like the Himalayas, and therefore to plate tectonics.

Plate Tectonics and Ocean Currents

Plate tectonic processes are also responsible for climate change in other ways. For example, the movement of plates can change the characteristics of ocean basins, and that can change ocean currents and, therefore, the climate. Prior to about 40 Ma, the southern end of South America was still connected to Antarctica, or at least the passage between them was too shallow to allow significant water flow. Sometime between 41 and 34 Ma, that body of water, the Drake Passage, was widened and deepened by plate motion, and since then the strong Antarctic Circumpolar Current has flowed around the continent in a west to east direction (figure 3.6).

This current has the effect of isolating Antarctica from relatively warm ocean currents of the southern Pacific, Atlantic, and Indian

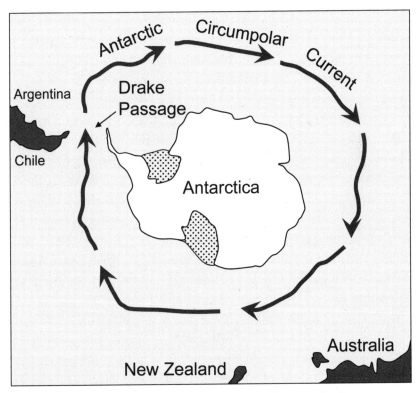

Figure 3.6. The approximate path of the Antarctic Circumpolar Current

Oceans. That has kept warm water away from Antarctica and is responsible for the glaciation of the southern continent starting at about 35 Ma and continuing until today (with a possible interruption between 25 and 15 Ma) (figure 3.5).

Between about 100 Ma and 10 Ma, North and South America were separated from each other by a waterway hundreds of kilometers wide; under those conditions, water was able to flow freely between the Pacific and Atlantic Oceans. But there was ongoing subduction of oceanic crust beneath what is now Central America. In a manner similar to what is shown on figure 3.2 between India and Asia, that process led to formation of magma above the subducting plate (not because the subducting plate was melting but because it was releasing water that was mixing with the overlying hot mantle rock, inducing that rock to melt by flux melting). That led to many millions of years of volcanic activity, and to the formation of a series of volcanic islands within what is now Central America (figure 3.7). Finally, at around 10 Ma, those volcanic islands coalesced into an isthmus that opened the way for land animals to pass between North and South America but blocked the Central American Seaway.

That change had the effect of making the Gulf Stream (and the entire Atlantic circulation system) more intense, and the warm water flowing north brought more warmth and more moisture to the northern Atlantic. Ironically, that additional warmth and moisture led to more intense snowfall in Iceland, Greenland, northern North America, and northern Europe, and thus to a lower albedo, and eventually to the beginning of the Pleistocene Glaciations.[10] The northern hemisphere has been repeatedly glaciated since 2.5 Ma, in cycles that have remarkably regular periodicity. The origin of those cycles is discussed in chapter 6.

Summary

The solar evolution process discussed in chapter 2 happened (and is still happening) on a time scale of billions of years. The various climate-changing processes related to plate tectonics have taken place a little faster—on time scales of hundreds of millions of years to tens of millions of years, to just millions of years. It took hundreds of millions of years for Rodinia to be assembled and moved into an Earth-cooling equatorial position, many tens of millions of years for the construction of the Himalayan range and then the ongoing erosion that dramatically cooled the planet, and millions to tens of millions of years for the changes to the Antarctic Circumpolar Current and the Isthmus of Panama to lead us into glaciations in Antarctica and then the northern hemisphere.

The climate impacts of these plate-tectonic processes are subtle and, on their own, not enough to produce the significant climate changes of the Cryogenian or the Cenozoic glaciations. In all of these

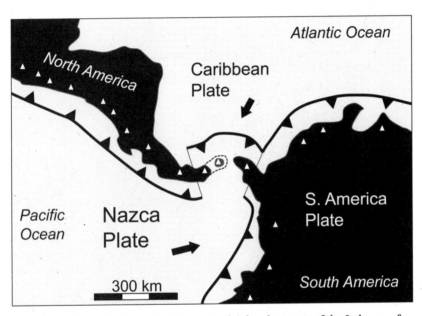

Figure 3.7. Scenario of the final stages in the development of the Isthmus of Panama. At around 15 Ma, the Nazca Plate was subducting (along the toothed lines) beneath North and South America, while a small part of the Caribbean Plate was subducting in the other direction beneath what is now Panama. The white triangles are possible locations of volcanoes. The dashed line shows the likely next step in the construction of the isthmus. By the author, based on León, S., et al., "Transition from Collisional to Subduction-related Regimes: An Example from Neogene Panama-Nazca-South America Interactions," *Tectonics*, V. 37, pp. 119–39, 2018.

cases, positive feedback mechanisms magnified the gentle forcing of the tectonic processes to produce global-average temperature changes ranging from several degrees to several tens of degrees (C). As we look at the approximately 1°C of warming that we have already caused, we need to remember that this is just the nudge, and that climate feedbacks are going to make the bigger difference.

4

COOLING AND WARMING FROM VOLCANIC ERUPTIONS

What, then, shall I say about the fruits of the earth that
year, when the weather was so remarkably unseasonable
that the warmth of the Sun was hardly able, even a
little, to reach the earth, and the fruits of that year could
barely attain maturity, if at all? For so great a thickness
of clouds covered the sky throughout that whole summer
that hardly anyone could tell whether it was summer or
autumn. The hay, drenched incessantly by strong rains
that year, was unable to dry out, because it could not
collect the warmth of the Sun on account of the thickness
of the clouds.

— Richerus of Sens, 1267[1]

THE BENEDICTINE MONK Richerus was referring to the dreadful
summer of 1258 as experienced in Sens, a village (at that time)
southeast of Paris. This type of climate chaos was recorded across Europe that year and was also felt in most other places of the world, although records are sparse from other regions. Because it was cold and wet throughout the summer of 1258 (and into 1259), there was widespread famine and sickness. Richerus, who likely had no idea that the miserable weather was related to a volcanic eruption—especially not one 2,240 leagues away[2]—doesn't seem to have suffered too badly, but the death rate was high amongst those without resources.

This chapter is about why we have volcanic eruptions, how they differ in style and in size and why that's important, and how they affect the Earth's climate both on a short time scale, as was the case in 1258 and 1259, and on a much longer time scale.

Volcanic Eruptions

Earth is a volcanic planet because the core and mantle are still hot, and because there is still convection in the mantle. There is reason to believe that the Earth was even more volcanically active in the distant past (billions of years ago, not millions) and that, in many cases, the erupting magmas were hotter than they are now. Venus is also still volcanically active—possibly even more than the Earth—but Mars and our Moon are both small enough to have cooled down to the point where they are no longer active.

Volcanism takes place in several different settings on Earth, and most are either directly or indirectly related to plate tectonics (figure 4.1). Before we go any further here, we need to talk a little about some of the parts of Earth that are relevant to plate tectonics and volcanism.

The uppermost layer of the Earth is the crust. It is brittle rock that is 30 to 40 km thick on the continents and 5 to 6 km thick under the oceans. Continental crust consists of relatively light-colored low-density rock, with an overall composition similar to that of the rock granite. Rock of this composition is described as *felsic* because of

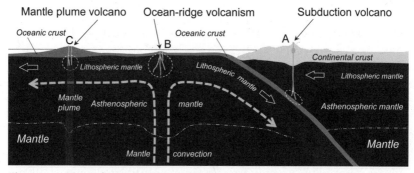

Figure 4.1. The more important settings of volcanism on Earth, shown in the context of plate tectonics. The diagram is a cross-section through the crust and into the upper part of the mantle, to a depth of about 400 km.

its high levels of the mineral feldspar and quartz (silica). Owing to the low density of felsic rock, the continents float a little higher on the mantle than does the ocean floor. Oceanic crust is darker and of greater density, with a composition similar to the rock basalt. This composition is known as *mafic* because of its higher levels of magnesium and iron. It floats a little lower on the mantle because of its greater density, and that's why it is covered with sea water almost everywhere.

The uppermost part of the mantle is rigid (hence the name "lithospheric mantle" or "rock-like mantle"). The lithospheric mantle and the crust make up the lithosphere, which is about 100 km thick in total, and is what the tectonic plates are made of. The next layer of the mantle is close to its melting point and so is known as the asthenosphere (meaning the "weak layer"). The rest of the mantle (all the way down to the boundary with the Earth's core, 2,900 km) is strong but plastic rock that is slowly (cm/year) convecting in response to the transfer of heat from the core.

A mantle plume is a column of hot mantle rock, not magma, that is rising up from near to the mantle/core boundary at a rate about ten times faster than the rate of mantle convection. Mantle plumes appear to be unaffected by mantle convection, and as far as we can tell, they tend to stay in one place for many tens of millions of years.

The magma that erupts at volcanoes is typically derived from one of three settings labelled A, B, and C on figure 4.1. Setting A is within the asthenosphere immediately above a body of subducting oceanic crust. Oceanic crust is both wet (has water in cracks and pores) and contains hydrous minerals.[3] When this crust gets heated up by being forced down into the hot mantle, the hydrous minerals lose their water, and the water released, along with water from the cracks and pores, rises up into the overlying asthenosphere, where it acts as a flux to reduce the melting temperature of the rock there. That results in partial melting of the already hot mantle rock (by flux melting) and production of magma, which rises toward the surface to form a subduction volcano. Mount St. Helens in Washington State is a good example of this type of volcano.

Setting B is at a mid-ocean spreading ridge. Hot mantle rock from depth is brought slowly toward surface by mantle convection, and the resulting drop in pressure leads to partial melting of that rock. This is known as decompression melting. The resulting mafic magma (rich in magnesium and iron) rises through cracks in the diverging plates and out onto the seafloor, where it forms new oceanic crust of basaltic composition. This process is happening all the way down the middle of the Atlantic Ocean and in many other areas.

Setting C is above a mantle plume. As the rock of the mantle plume rises toward surface, the same decompression melting takes place, and mafic magma is produced. At first it forms basaltic rock on the seafloor, but if the process continues for long enough, an ocean island will form. Kilauea, on the island of Hawaii, is a good example of this type of volcano.

The magma that forms at spreading ridges and mantle plumes (settings B and C) starts out with a mafic composition, like that of basalt, and stays that way because there is limited opportunity for it to change. In contrast, the magma at a subduction zone (setting A) also starts out with a mafic composition, but because it travels a long way through the crust and tends to get stored in a magma chamber within the crust, it stands a good chance of changing along the way (figure 4.2). First, the heat of the magma leads to partial melting of the surrounding crustal rock, and that adds silicon to the magma, making the magma composition less mafic and more felsic. Second, there is the opportunity for crystals to form in the magma chamber, and those will be crystals of minerals that are rich in iron and magnesium and poor in silicon. As they settle to the bottom of the magma chamber, they will contribute to the zonation of the magma chamber. The magma at the top will be more felsic than it was, and the magma at the bottom will be more mafic because those crystals will remelt at the bottom, where it is hotter.

The point of all of this is that the more felsic magma that typically erupts at a volcano associated with subduction tends to be more viscous (less runny) than the mafic magma that erupts at a spreading ridge or a mantle plume. That greater viscosity typically makes it

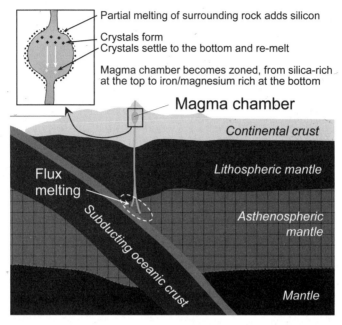

Figure 4.2. A subduction zone volcanic setting. The processes that lead to modification of the magma into a more silica-rich type are shown in the inset.

erupt less frequently, but also more explosively when it does, sending massive amounts of ash and gases far up into the atmosphere (as in figure 4.3).

Volcanic Eruption Products

The products of volcanic eruptions include flowing lava (magma), volcanic ash (tiny shards of glass) and rock fragments (collectively known as tephra), and gases, but the proportions of these vary quite widely depending on the type of eruption.

An explosive eruption (which is common at a volcano above a subduction zone, especially when the magma is felsic) emits a large volume of tephra, and if it's a big eruption, this material is forced well up into the stratosphere. The 1980 eruption of Mount St. Helens was a moderately large event during which the volcanic ash column extended to an elevation of about 24 km (figure 4.3). However, almost no lava flowed during the 1980 eruption.

If the erupting magma has a mafic composition, it tends to be runny, and so will likely flow out relatively gently as lava, known as an effusive eruption, with little or no volcanic ash emitted.

Volcanic ash particles are small, but they are not so small that they stay aloft for a very long time. The larger particles come down in hours to days, and the smaller ones in days to weeks. So, while a thick cloud of ash can block the sun for a while, and provide a little cooling, it doesn't typically last long enough to have a climate impact. And where the ash settles onto ice or snow, it has a warming effect (because of the reduced albedo) that could cancel out some of the cooling

Figure 4.3. Part of the volcanic ash column of the May 18, 1980, eruption of Mount St Helens. Photo by U.S. Geological Survey, public domain.

effect. Volcanic gases, on the other hand, do have a significant climate impact.

Whether it is an explosive or an effusive eruption, or a combination of the two, a large volume of gas is always released during a volcanic event. Water vapor is the most abundant volcanic gas, but the amount released by volcanic eruptions is very small compared with the amount of water already in the atmosphere and so doesn't have climate implications. From the perspective of climate change, carbon dioxide and sulphur dioxide are the most important volcanic gases. Carbon dioxide is a greenhouse gas, of course, and so volcanism can lead to warming. Sulphur dioxide does have the properties of a greenhouse gas (it absorbs infrared radiation), but it doesn't last long in the atmosphere. Instead, within hours to days, it reacts with water to become sulphuric acid droplets or with calcium and oxygen to become tiny calcium sulphate crystals called sulphate aerosols. Being tiny, they can remain in the atmosphere for a period of months to several years. They block the sun, and as Richerus observed, they can have a strong cooling effect.

To understand the significance of volcanic gases, we need to consider the amounts of gas emitted in a typical eruption compared with the amounts typically present in the atmosphere (table 4.1). It's easy

Table 4.1. Current atmospheric reservoirs of some volcanic gases, compared with amounts emitted during a typical large eruption. H_2O and CO_2 reservoir numbers are derived from atmospheric concentrations. The sulphur reservoir value is from Manktelow, P., et al., "Regional and Global Trends in Sulfate Aerosol Since the 1980s," *Geoph. Res. Letters*, V. 34, L14803, 2007. The data for the 1991 Pinatubo eruption are from Self, S., et al., "The Atmospheric Impact of the 1991 Pinatubo Eruption, 1997," pubs.usgs.gov/pinatubo/prelim.html.

	H_2O	CO_2	$SO_4 + SO_2$
Current atmospheric reservoir (millions of tonnes)	16,000,000	3,200,000	2
Amount emitted during the 1991 Pinatubo eruption (millions of tonnes)	400	40	20

to see that the amount of water emitted by a moderately large eruption, such as Pinatubo in 1991, is insignificant compared with the amount already in the atmosphere. It's also important to recognize that water has a relatively short lifetime in the atmosphere (about 9 days), so the 400 million tonnes added by Pinatubo was likely rained out within a week or two and would have had no climate impact.

Volcanic carbon dioxide emissions are also very small compared with the current atmospheric reservoir, but CO_2 has a much longer residence time (hundreds to thousands of years),[4] so there is the potential for volcanic CO_2 to lead to warming if a higher-than-average level of volcanism is sustained for centuries or more.

On the other hand, volcanic sulphur emissions are large compared with the atmospheric reservoir of sulphur, and that's why a large volcanic eruption can have a rapid and significant climate effect (cooling, because sulphate aerosols block incoming sunlight). As already noted, sulphate aerosols do not stay in the atmosphere for more than a few years in most cases, so the climate effect tends to be quite short.

The Climate Effects of Past Volcanic Eruptions

Some notable volcanic eruptions, and what we know about their climate impacts, are described here. In each case, the estimated volume of magma erupted, either as flowing lava or as tephra (ash and pumice), is given. For reference, the well-known 1980 eruption of Mount St. Helens, Washington State, lasted for nine hours, during which time 0.21 km³ of tephra was ejected into the upper troposphere and lower stratosphere.

Kilauea, Hawaii

Hawaii's Kilauea volcano is a mantle plume volcano that started forming about 300,000 years ago, although the mantle plume is more than 80 million years old. Kilauea has been frequently active for most of those 300,000 years, and during historical times, it has been erupting more often than it has been quiet. The most recent eruption cycle started in 1983 and continued, almost without interruption, until

Figure 4.4. A gas monitor on the floor of the Kilauea Caldera. The white clouds are emitted water vapor, but there is also a strong smell of SO_2 at this location.

2018. Over those 35 years, approximately 4.4 km³ of magma erupted, mostly as lava, with a small amount of tephra in the early stages.

There is no evidence that this recent eruption cycle has had a measurable impact on the global climate. The average rate of magma output over the 35 years was low (about 0.01 km³/month), and while a large volume of gas was emitted (figure 4.4), it wasn't enough to make a significant difference to the atmosphere as a whole.

Pinatubo, Philippines

As shown in table 4.1, the major eruption of Mt. Pinatubo (Luzon Island, Philippines) in June 1991, produced a significant spike in the amount of sulphate in the atmosphere. Like the 1984 Mount St. Helens eruption, the Pinatubo eruption lasted for about nine hours, but in that time, almost 25 times as much tephra was ejected (5 km³) as at

Mount St. Helens, and correspondingly higher volumes of gas were released.

The atmospheric effect of volcanic eruptions between 1960 and 2020 is depicted on figure 4.5. This shows the degree to which solar radiation is blocked by the atmosphere, resulting in dimming of sunlight, most of which can be attributed to sulphate aerosols from major volcanic eruptions. Events that had a significant insolation dimming effect are the 1963 to 1964 eruption of Mt. Agung on Bali, Indonesia (several separate eruptions over about 15 months), the 1982 eruption of El Chichón, Chiapas State, southern Mexico (with just over 2 km³ of volcanic ash emitted), and the June, 1991 eruption of Pinatubo. (Note that the 1980 eruption of Mount St. Helens doesn't even register.) Although El Chichón resulted in greater dimming, significant dimming from Pinatubo lasted a little longer. The inset in figure 4.5 shows that the dimming extended for about two years but was minimal after about 18 months. That amount of insolation dimming resulted in a

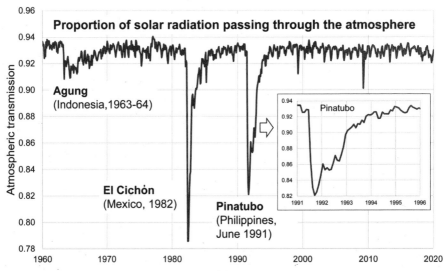

Figure 4.5. Evidence of atmospheric sulphate aerosol levels associated with major volcanic eruptions from 1960 to 2020, based on solar radiation dimming. The inset shows the dimming associated with the 1991 Pinatubo eruption in a little more detail. Based on data from NOAA Earth System Research laboratory, Global Monitoring Division, https://www.esrl.noaa.gov/gmd/grad/mloapt.html

maximum global temperature drop of about 0.5°C and in measurable cooling from June of 1991 until 1993.

Laki, Iceland

Volcanism on Iceland is related both to the mid-Atlantic divergent plate boundary and to a mantle plume. Iceland exists solely because of that volcanic activity, which is typically mafic in composition and leads to effusive (lava flow) eruptions that are less dangerous than explosive eruptions.

The largest Icelandic eruption in historical times, at Laki in the south-central part of the country, started in June 1783 and continued until February 1784.[5] Over that eight-month period a total of 14 km^3 of lava erupted, at a flow volume roughly 175 times greater than the 1983 to 2018 eruption at Kilauea. Most of that volume was flowing lava, but there were frequent explosive eruptions, especially in the early months. The volume of SO_2 released was about six times greater than that of the 1991 Pinatubo eruption, partly because the eruption volume was greater but mostly because mafic magmas have higher sulphur levels than felsic magmas. Most of the sulphur (~80%) was released by the explosive events.

The Laki eruption was devastating in Iceland. There was the strong cooling of the climate due to sulphate aerosols, but more serious was that, along with other gases, a significant volume of hydrofluoric acid was released and spread across the country. This was ingested by grazing livestock and led to the death of approximately 60% of farm animals. The ensuing famine (from this and other causes) resulted in the death of about 9,000 Icelanders, or 20% of the population.

The Laki eruption had significant climate effects outside of Iceland, mostly in the northern hemisphere. As shown on figure 4.6, there was a temperature drop in Europe and North America of more than 1.0°C for over two years, and of at least 0.5°C for close to four years. It is reported that the Mississippi River froze over at New Orleans, that changes to monsoon patterns brought drought and famine to many parts of Africa and Asia, and that the famine in Europe may have been a factor in the onset of the French Revolution of 1789.[6]

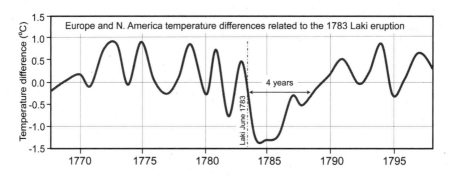

Figure 4.6. Annual temperature differences from average in Europe and North America from 1768 to 1798[7]

Samalas, Indonesia

Indonesia's Samalas volcano is related to subduction; hence the magma is typically felsic, and the eruptions are normally explosive. The 1257 eruption was certainly that, and it is also considered to represent the largest source of volcanic gases in historical times.[8] It is estimated that approximately 10 km³ of magma was released along with about 160 million tonnes of SO_2 (compared with ~120 million tonnes for Laki and 20 million tonnes for Pinatubo). A critical factor concerning the short-term climate effect is that the release of gases at Samalas likely took place over roughly a single day, while at Laki it was several months.

As we've heard already from Richerus, there was strong cooling in the northern hemisphere in 1258 and for a few years after that, but we don't have much information relating to other parts of the world from that time, nor sufficient global temperature data, to assess the global impact of the Samalas eruption.

Toba, Indonesia

Although there have been many large volcanic eruptions in the past several tens of thousands of years, historical information is lacking, or extremely sparse, for any prior to 1257. For most older eruptions, we have to rely on geological, archaeological, and paleontological records, and those don't tend to be as definitive as written records based on direct observations.

The now extinct Toba volcano is situated in Sumatra, Indonesia, above the same subduction zone that was the site of the devastating earthquake and tsunami in December, 2004. There is clear geological evidence for a major eruption at Toba about 74,000 years ago, including a massive crater (figure 4.7), and very extensive and thick ash layers across much of southern Asia. The volume of the erupted material was massive, in the order of 3,000 km^3 (~300 times that of Samalas), and it is estimated that 6 billion tonnes of SO_2 was released (~50 times that of the Samalas eruption).[9]

There is no question that the Toba eruption had a significant climate impact, although it's not that easy to determine what that impact was. A study of ice cores from different locations on Greenland and Antarctica has shown evidence of widespread elevated sulphate aerosol levels for at least a decade and isotopic evidence for a

Figure 4.7. Lake Toba, Sumatra, Indonesia, lies within a crater that is about 90 km long and 40 km wide. Image from NASA, public domain.

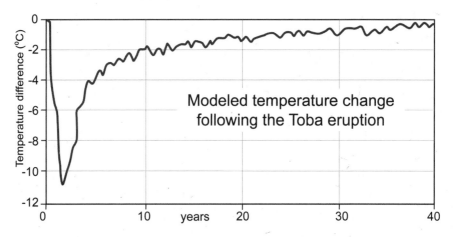

Figure 4.8. Modeled global temperature change effect of the 74,000-year-old Toba eruption. Based on a figure in Robock et al., "Did the Toba Volcanic Eruption of ~74k BP Produce Widespread Glaciation?" *Journal of Geophysical Research*, V. 114 (D10), D10107, 2009

temperature drop.[10] The results of modeling of the climate effect of the Toba eruption, based on assumptions about the volume of SO_2 emitted, are shown on figure 4.8. It is estimated that the global temperature dropped by over 10°C shortly after the eruption, remained at least 4°C colder than normal for about 4 years, at least 2°C colder for nearly 10 years, and 1°C colder than normal for almost 30 years. But there is no evidence from this modeling, or from the ice-core records referred to above, that the Toba eruption pushed the Earth into a glacial period.

One of the theories surrounding the Toba eruption is that it represented a significant threat to early human populations and brought *Homo sapiens* close to extinction. Although that is an interesting concept, there is no evidence that human populations were decimated by the few years of extreme cold and the few decades of cool weather.

Yellowstone, Wyoming

Volcanism in the Yellowstone region has long been considered to be the product of a mantle plume that used to be under southern Oregon, and then southern Idaho, and now northwestern Wyoming—

not because the mantle plume has moved but because the North America plate has slowly moved toward the southwest, over top of it. There is now some question about this hypothesis, but that doesn't need to concern us here. If the source is a mantle plume, the process is different than at other mantle plumes, such as Hawaii, Iceland, or Galapagos, because, while those volcanoes have consistently mafic magma, the magma at Yellowstone is typically felsic, and most large eruptions have been explosive. The ultimate source magma is almost certainly mafic, so some significant differentiation of the magma must have taken place to make it felsic, likely involving partial melting of crustal rocks to add more silica to the magma.

The last major eruption at Yellowstone was 639,000 years ago. It was huge, with about 1,000 km^3 of tephra erupted, and the ash covered a good part of the continent, east as far as Mississippi, west to the Pacific Ocean, north into southern Canada, and south into northern Mexico.

A recent study of ocean sediments in the Santa Barbara Basin, offshore southern California, has revealed two layers of Yellowstone ash from that time, suggesting that there were two separate eruptions.[11] Along with the ash layers are the shells of planktonic marine organisms that can be used to estimate water temperatures. These results show that both eruptions led to ocean surface water temperature declines of approximately 3°C in that area, in each case for several decades. That result is highly significant because, under these conditions, land areas typically cool several times more than ocean water.

Deccan Traps, India

The Deccan Traps basaltic lava flows cover an area of west-central India exceeding 500,000 km^2 (almost as big as Texas) in places up to 2,000 m thick. The volume of lava still remaining (i.e., not eroded over the past 66 million years) is around 1 million km^3, but it is likely that the initial volume was at least twice that. Most of this lava erupted over a period of about 35,000 years close to 66 Ma. To put that into context, the rate of eruption during this period at Deccan was about 14 times the rate of the 8-month Laki eruption or 2,500 times the rate of the recent 35-year Kilauea eruption. Deccan is one of several extraordinary volcanic deposits around the world known as Large Igneous Provinces (LIP). Another example, closer to home, is the Columbia River Flood Basalts, which covered an area of about 160,000 km² across Washington, Oregon, and Idaho with up to 3,500 m of lava between 17 and 15.5 Ma. All LIPs are assumed to be related to mantle plumes, but as is evident from the comparison with Laki and Kilauea, these are mantle plumes on steroids!

While it was active, the Deccan eruption would have caused strong cooling due to sulphate aerosols, but the large amount of CO_2 emitted over thousands of years (estimated at 70 trillion tonnes[12]) made a significant difference to the concentration of CO_2 in the atmosphere and, therefore, led to warming that would have lasted for much longer. The modeled extent of that warming is shown on figure 4.9. The

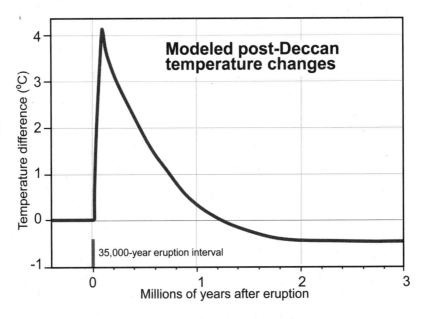

Figure 4.9. Modeled global temperature change effect of the 66 Ma Deccan Traps eruption. Based on a figure in Dessert et al., 2001. Those authors did not consider the cooling effect of SO_2 released by the Deccan eruptions, but it is likely that there was significant short-term cooling (decades to centuries) associated with each major pulse of magma at Deccan.

authors of that study estimate that the Earth's temperature would have increased by around 4°C for a few tens of thousands of years and by at least 2°C for about 400,000 years. But after 1.2 million years, there would have been cooling of about 0.5°C (relative to pre-eruption times) because weathering of the basalt would have consumed a lot of atmospheric CO_2 (see chapter 3).

The end of the Cretaceous Period at 66 Ma coincides with the extinction of the dinosaurs and many other organisms—in fact, about 75% of species—and for many decades, it was widely believed that the Deccan eruption was the main cause of that massive extinction. That view changed in the early 1980s with the well-known impact theory for the end-Cretaceous extinction (more on that in chapter 8), and now most geoscientists agree that the impact was probably the main cause. But an interesting twist to the story is that there is evidence that a major increase in the rate of volcanism (represented by the 35,000-year event on figure 4.9) coincides closely with, and may have been triggered by, the Chicxulub impact on the far side of the world (Yucatan, Mexico).[13] In other words, it is possible that both the Chicxulub impact and the Deccan volcanism played a role in the devastation at the end of the Cretaceous.

Siberian Traps, Russia

The Deccan Traps eruption was large, but the Siberian Traps eruption of 252 Ma was about four times larger. The volume of magma is estimated to be 4 million km^3, and the flows covered an area about the size of Australia. It is estimated that about two-thirds of the magma erupted over approximately 300,000 years just prior to the end of the Permian Period (251.9 Ma),[14] and it coincides with the most catastrophic extinction of all time. Over 95% of all marine species and 70% of terrestrial species disappeared from the fossil record at the end of the Permian. Life on Earth was forever changed, or put differently, the future course of evolution—including the origin and evolution of mammals—was significantly affected by that event.

There is isotopic evidence for strong warming, in the order of 10°C, for at least the first 10 million years of the Triassic. Figure 4.10

provides a concept of how climate change might have progressed from the start of the Siberian Traps eruption (time 0) and then for the first million years of the Triassic. Assuming that the volcanic eruption proceeded episodically (which is typical), there would have been short periods of cooling caused by sulphate aerosol pulses and increasingly intense warming caused by the progressive buildup of atmospheric carbon dioxide.

The authors of this study didn't include the potential effects of increased weathering in their model, but it's likely that there was such an effect, so the temperature may have eventually dropped below the pre-eruption level.

While much of the end-Permian climate crisis and extinction might have been due to the significant volume of volcanic gases re-

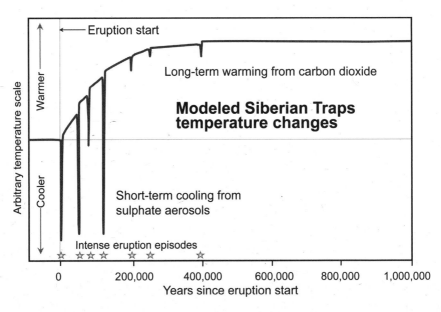

Figure 4.10. Conceptual model of global temperature changes for one million years following the start of the Siberian Traps eruption. The stars represent possible individual episodes of intense volcanism. Based on a figure in Black, B., et al., "Systemic Swings in End-Permian Climate from Siberian Traps Carbon and Sulfur Outgassing," *Nature Geosci*, V. 11, pp. 949–54, 2018.

leased in Siberia, there were probably some strong positive feedbacks that enhanced and also extended the warming, including the likelihood that massive amounts of methane were released from seafloor methane hydrate deposits.

Volcanism Versus Anthropogenic Effects

It is clear from the foregoing that volcanic eruptions can have significant implications for our climate and for life on Earth. Sulphur emitted by large eruptions, such as Samalas in 1257, typically leads to

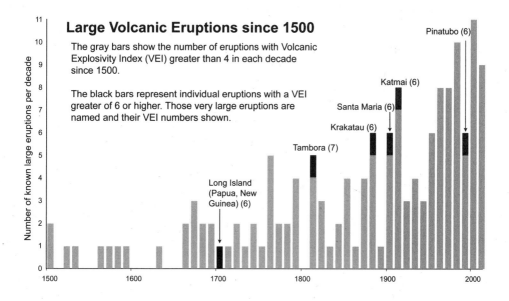

Figure 4.11. A summary of known large volcanic eruptions over the past 520 years. Based on Bradley, R., and Jones, P., "Records of Explosive Volcanoes over the Last 500 Years," in Bradley, R., and Jones, P., (eds), *Climate Since 1500*, Routledge, London, 1992; and *Wikipedia* for the period 1992 to 2020. Volcanic Explosivity Index is based partly on the explosivity of an eruption and on the volume of magma or tephra released. A VEI of 4 is roughly equivalent to an eruption volume greater than 0.1 km^3. A VEI of 6 represents an eruption volume of greater than 10 km^3. volcanoes.usgs.gov/vsc/glossary/vei.html. Each of the very large eruptions (VEI 6 or higher) had emissions that were many times greater than all other smaller eruptions in each century.

cooling, lasting for months or years—or even decades as in the case of Toba. The carbon dioxide emitted by eruptions of this size tends to be too little, compared with the atmospheric reservoir, to have a climate impact.

On the other hand, very large and sustained eruptions, such as the Deccan or Siberian Traps, can lead to significant long-term warming, and the effects of that warming can be catastrophic for life.

One of the common arguments against taking action on climate change is that the climate changes we've experienced can be ascribed to volcanism, as opposed to anthropogenic (human-caused) GHG emissions. There are several reasons why this is argument is not valid:

- Typical volcanic eruptions do emit CO_2, but the amount released is very small compared with the emissions from burning fossil fuels. In 2019 anthropogenic CO_2 emissions totaled nearly 37 billion tonnes. The U.S. Geological Survey estimates that all eruptions, both on land and beneath the sea, in a typical year emit about 200 million tonnes of CO_2, or about 0.5% of the amount emitted by burning fossil fuels.

- Even the largest eruptions in historical times have led only to cooling, not warming. It is only massive eruptions like Deccan that have led to warming, and there hasn't been an eruption on that scale since the Columbia River basalts 17 million years ago.

- There is no evidence for a reduction of volcanic activity that might be used as an argument that there has been less volcanic cooling happening in recent decades. Figure 4.11 shows an apparent increase in activity over the past five centuries, but this is likely an artifact of detection and reporting.[15]

5

EARTH'S ORBITAL VARIATIONS

Once you catch a big fish, you can't be bothered with small ones any longer. I worked for 25 years on my theory of solar radiation and now that it's complete, I'm without work. Theories of the magnitude of the one I have completed do not grow on trees, and I am too old to start a new theory.

— Milutin Milanković in 1941 (age 62)[1]

THE BIG FISH THAT Milutin Milanković referred to in 1941 was the completion of his book *Canon of Insolation and the Ice-Age Problem* in which he argues how the natural variations in the shape of the Earth's orbit around the sun and in the tilt of the Earth's rotational axis played a critical role in the timing of glaciations over the past two million years. Milanković wasn't the one who discovered that the Earth's orbit and tilt changed over time; that was the work of Hipparchus of Nicea way back in 130 BC and of Johannes Kepler in 1609. Nor was he the first to speculate that these changes might affect the climate; that can be ascribed to the Scottish scientist James Croll in 1875. But Milanković was the first to calculate the effects of insolation at different latitudes and to accurately determine the periods during which those changes would be most likely to contribute to the growth of glaciers and to the shrinkage of glaciers.

Changes in the Earth's Orbit and Tilt

The Earth's orbit and tilt variations are not easy to understand, and nor is it easy to grasp how they affect the climate, but understanding these changes is critical to our understanding of past (and future) climate changes, and this is going to come up again in later chapters, so let's give it a shot.

The first important variation is in the shape of the Earth's orbit around the sun and the position of the sun within that shape. As shown on figure 5.1, the Earth's orbit is not a circle; it is an ellipse.[2] More importantly, the sun is not at the center of the ellipse. This means that at different times of the year the Earth is closer to the sun than it is at other times.

Not only is the Earth's orbit elliptical, the shape of the ellipse changes with time. On a consistent cycle of approximately 100,000 years, the shape changes from just a little bit elliptical to a bit more elliptical. This is illustrated on figure 5.2. When the orbit is more elliptical, the eccentricity of the sun's position within the orbit is also more extreme, and so the difference between the minimum and maximum Earth–sun distances is higher. The degree of eccentricity has an effect on our climate, and so, on a time scale of 100,000 years, that effect increases and decreases.

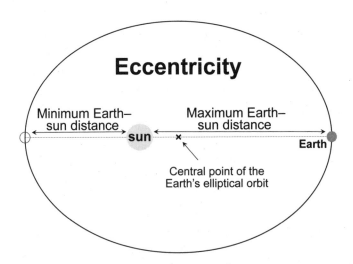

Figure 5.1. A representation of the elliptical nature of Earth's orbit around the sun, looking down from above. The orbit is an ellipse; the x marks the center of the ellipse; and the sun is off-center (i.e., eccentric). The eccentricity is exaggerated to make the point.

The second important feature of Earth's motion is the tilt of the rotational axis (also known as obliquity) relative to the plane of the orbit around the sun (figure 5.3). At present our axis is tilted at 23.5° from "vertical," but that varies from 22.1° to 24.5° on a time scale of close to 41,000 years.

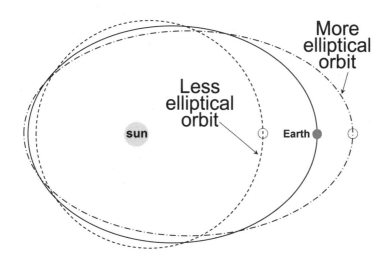

Figure 5.2. A representation of the variations in the elliptical nature of Earth's orbit around the sun. The elliptical shapes of the orbits are greatly exaggerated in this diagram.

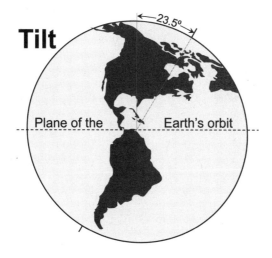

Figure 5.3. A depiction of the tilt of the Earth's axis

The tilt of the Earth's rotational axis is what gives us seasons (figure 5.4). We have summer in the northern hemisphere when the Earth is in the part of its orbit where the northern hemisphere is pointed toward the sun, and summer in the southern hemisphere when the southern hemisphere is pointed toward the sun. At times of greater tilt, the seasons are slightly more exaggerated than they are now (colder winters, hotter summers). At times of lesser tilt, they are less exaggerated (warmer winters, cooler summers).

The third aspect of the Earth to consider is the variation in the direction of tilt (also known as precession). The Earth's spin tends to keep it pointing in the same direction (just as a gyroscope tends to be stable), but in fact that direction is very slowly changing (wobbling if you like), such that, in 23,000 years from now, it will be pointing in the opposite direction.

If all of this has your head spinning (on an angle that changes), don't panic; we're going to make some sense of it in the next few paragraphs.

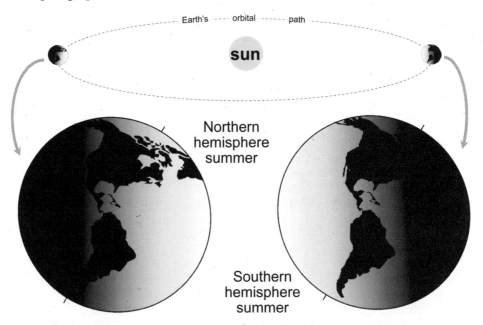

Figure 5.4. The effect of the Earth's tilt on seasons, showing a northern hemisphere summer on the left and a southern hemisphere summer on the right.

Milanković Cycles and Climate

All of the cyclical variations that are described above have implications for our climate (and therefore also for glaciation). Although from year to year the amount of energy the *whole* Earth receives from the sun doesn't change at all, there are changes in where on Earth that solar energy is strongest and at what time of the year it is strongest. Milanković and some colleagues, including Alfred Wegener and Wegener's father-in-law Wladimir Koppen, realized, for example, that glaciers grow best at temperate latitudes—in fact at around 65° north or 65° south[3]—and that they can start growing only on land. As shown on figure 5.5, 65° north passes through Alaska, northern Canada, Greenland, Iceland, Scandinavia, and Russia. It's pretty much land the whole way. On the other hand, 65° south is entirely in the Southern Ocean. There is almost no land at all, and so there is very little chance for glaciers to start forming in that area.

Based on this information, Milanković decided that insolation variations at 65° north were what mattered the most, and so he calculated the variations at that latitude. He also focused on insolation in summer because he was aware that cool summers are more important than cold winters to the growth of glaciers. That may be counterintuitive, but it's because less snow melt when the summers are

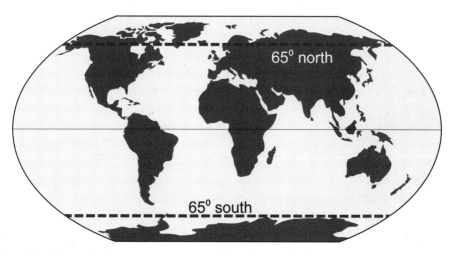

Figure 5.5. Land versus sea at 65° north and at 65° south

cool and also because very cold winters tend to be drier than warmer winters and so less snow falls.

The climate and glaciation effects of the three orbital parameters—eccentricity, tilt, and tilt direction—are summarized in table 5.1.

As already noted, Milanković published his major research work in 1941, right in the middle of WWII. Alas, few of his contemporaries who thought and wrote about glaciation and glacial cycles gave it much credence, and for the next 35 years, his "big fish" was a really just a big flop.

Like Alfred Wegener, Milutin Milanković had a theory so far ahead of its time that there wasn't enough evidence to demonstrate that it was reasonable. The first problem was that although it was widely accepted that glaciers had come and gone several times during the past million years, the timing of those events wasn't well known. So, while Milanković was able to use his theory to estimate the timing of past glaciations, it wasn't easy to verify his estimates. Another problem was that the calculated differences in forcing would not have been enough to drive glacial cycles. It turns out that the sceptics didn't

Table 5.1. The effect of Milanković cycles on the Earth's climate and on glaciation

Variation	Effect
Eccentricity (100,000-year cycles)	Eccentricity controls the variations in Earth–sun distances, and so in conjunction with the tilt direction, it determines how close or far the Earth is during the northern hemisphere summer. A high eccentricity provides a greater opportunity for the Earth to be pushed from a non-glacial state to a glacial state or vice versa.
Tilt angle (41,000-year cycles)	A greater tilt angle exaggerates seasonal differences. A lesser tilt angle leads to cooler summers and warmer winters, and that favors the growth of glaciers.
Tilt direction (23,000-year cycles)	Tilt direction is the key because it determines which hemisphere (north or south) is pointing toward the sun when the Earth is farthest away from the sun. Glaciation is favored when the Earth–sun distance is greatest during the northern hemisphere summer, leading to cool summers with less melting.

recognize the importance of climate feedbacks in amplifying the weak forcing of insolation differences. Like so many brilliant people, Milanković died (in 1958) without tasting the satisfaction of having his really important theory—his "big fish"—recognized by the rest of the science community.

Variations in the amount of solar energy received at 65° north during July for the past 250,000 years are shown on figure 5.6. The timing of the last two glacial periods, and of the last three interglacials (including the present one), are also shown on this diagram. Milanković and his contemporaries did not know the timing of the glaciations, but we do, and we can see reasonably clearly that the interglacials (warm periods) are related to strong peaks in summertime insolation at 65° north (peaks A, C, and E on figure 5.6). It's also evident that the last two glacial periods were jump-started by very strong

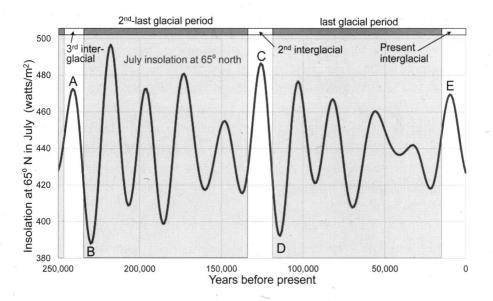

Figure 5.6. Insolation levels in July at 65° north for the past 250,000 years. Using data from Berger, A., and Loutre, M-F., "Insolation Values for the Climate of the Last 10 Million of Years, *Quaternary Science Reviews*, V. 10(4), pp. 297–317, 1991 (Supplement: Parameters of the Earth's Orbit for the Last 5 Million Years in 1 kyr Resolution).

minimums of summertime insolation at 65° north (troughs B and D on figure 5.6). But, apart from those correspondences, the insolation pattern is quite variable, and it's difficult to see how that variability is related to the two long periods of glaciation.

Over the several decades following the end of WWII, marine scientists and geologists started drilling into the soft sediments on the seafloor. The cores from these projects provided a wealth of information about past marine life and conditions of sedimentation. Eventually scientists started using isotopic methods to understand the conditions in which past marine organisms had lived, including the temperature of the water. A key paper, published in 1976, clearly showed the relationship between those temperature variations and the astronomical cycles. The authors wrote: "It is concluded that changes in the earth's orbital geometry are the fundamental cause of the succession of Quaternary ice ages."[4] This was the turning point for the Milanković concept. Since then, thousands of studies of climate variations have corroborated the key role played by the Milanković cycles, both during the Quaternary glaciations and in other climate cycles (e.g., Monsoons) for millions of years before then.

It's important to be clear that the Milanković cycles did not *cause* the Quaternary glaciations. Instead, it was the long slow decline in global temperatures of the past 50 million years, as described in chapter 3. That long period of cooling brought the Earth to the point where glaciation was possible; since then, the Milanković cycles have been the pacemaker of the glacial cycles.

Milanković Cycles Recorded in Ice

In the 1970s, a consortium of glaciologists from several countries started drilling through the thick glacial ice of Greenland and Antarctica. Over the next few decades, they recovered ice cores from holes extending down thousands of meters, retrieving samples of ice that had been laid down up to hundreds of thousands of years earlier. These cores not only allowed isotopic estimates of the temperatures at the time but also provided actual samples of the atmosphere (locked in bubbles in the ice) as it existed when the ice formed. And, unlike the deep-sea sediment cores, the ice cores have well-defined annual layers, so they can be time calibrated accurately (figure 5.7).[5]

Temperature data from an Antarctic ice-core are shown on figure 5.8, along with the pattern of July insolation at 65° north. These are temperature estimates based on measurement of the ratios of the hydrogen isotopes in the water molecules that make up the ice, and they represent the average annual surface temperature at the drill site over the past 250,000 years. The correlation between the temperature record and the July insolation levels is reasonably clear. The third-last interglacial, extending from 245 to 235 ka (245,000 to 235,000 years ago), corresponds with a period of high insolation. The following very low insolation initiated the beginning of the second-last glacial period. That was followed by a very high insolation period (at around 220 ka), which led to significant warming but wasn't enough to break the glacial cycle. Glacial conditions then intensified over the next 90,000 years.

Another period of very high insolation, culminating at around 120 ka, was able to break the cycle, leading to the second interglacial, which lasted from about 127 to 116 ka. That was followed by a similar cycle of increasingly cold climates and strong glaciation until around 20 ka, when the glacial cycle was again broken by a period of strong insolation.

Methane levels in air bubbles in the ice from the same core are shown on figure 5.9. These variations are even more closely correlated with the insolation pattern than are the temperature variations. Contrary to what you might think, methane variations were not *controlling* the climate over this time period; instead, they were *responding* to climate change. During periods of cooling, methane was stored

1836 m   1837 m

Figure 5.7. An ice core from a depth of 1,836 to 1,837 meters. Image from NASA, public domain.

in soil and permafrost, and during periods of warming and melting, methane was released (as it is being released right now because of anthropogenic climate change). Almost all of the insolation peaks throughout the 250,000-year period shown correspond with peaks in the atmospheric methane record.

Methane isn't the only positive feedback process that amplifies Milanković forcing. There is also a close correspondence between

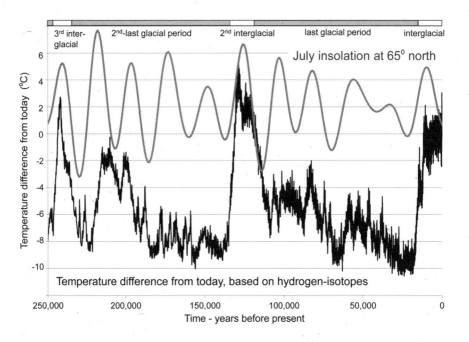

Figure 5.8. Insolation levels and Antarctic temperature differences from current for the past 250,000 years. Refer to figure 5.6 for the insolation units. Based on temperature data from the European Program for Ice-coring in Antarctica (EPICA) borehole at Dome C, Antarctica. The present-day average temperature at this site is about -50°C. This core penetrated through ice as old as 800,000 years, currently the longest ice-core record. Beyond EPICA, a project to drill in an area with ice that could be 1 million years old is in progress. Data from J. Jouzel, "EPICA Dome C Ice Cores Deuterium Data," IGBP PAGES, World Data Center for Paleoclimatology, Data Contribution Series # 2004 – 038, NOAA/NGDC Paleoclimatology Program, Boulder, CO, 2004.

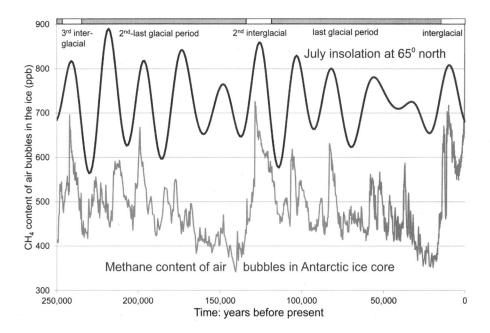

Figure 5.9. Insolation levels and Antarctic ice-core methane levels for the past 250,000 years (Data source: See Figure 5.8)

carbon dioxide levels and insolation values. The main driver of this feedback is the solubility of carbon dioxide in ocean water. As the climate warms, that solubility decreases, so more carbon dioxide comes out of solution into the atmosphere. Another strong feedback is the accumulation of ice and snow during cooling episodes, which increases the albedo of the planet, resulting in more solar energy being reflected away.

Milanković Cycles into the Future

Because the Earth's orbital cycles are well understood (thanks to Milanković), we can calculate insolation levels for any latitude and for any date in the distant past or the distant future. July insolation levels at 65° north for the past 150,000 years and the coming 150,000 years are shown on figure 5.10. We are heading into a long period of low ellipticity in the Earth's orbit (ellipticity is not shown separately on figure 5.10), and that's why the insolation levels won't vary a lot for the next 50,000 years. Numerous climate scientists have used this information to argue that the Earth is going to be in an interglacial climate for a long time to come.[6] Even after 50,000 years, possibly as far in the future as 100,000 years, there will be no very low or very high insolation events of the magnitude that pushed the Earth into and then out of glaciations many times over the past million years.

Some climate-change skeptics argue that anthropogenic climate warming is a good thing because it will keep us from plunging into the next glaciation, but this simply doesn't hold water, or doesn't cut any ice, if you prefer. Milanković cycles show us that the next glaciation is likely at least 50,000 years away, possibly 100,000. In fact, there is no natural process that we can anticipate happening that will help us out of the climate crisis. As we'll see in chapter 7, sunspot

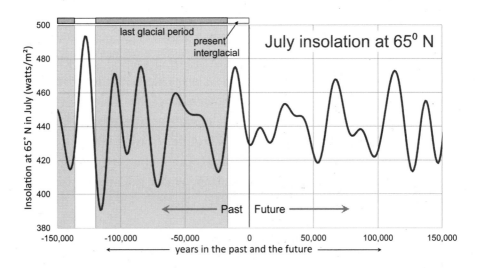

Figure 5.10. July insolation levels at 65° north for the past 150,000 years and the next 150,000 years. Based on data from Laskar, J., et al., "A Long-Term Numerical Solution for the Insolation Quantities of the Earth," *Astronomy and Astrophysics*, V. 428, pp. 261–85, 2004, vo.imcce.fr/insola/earth/online/earth /earth.html.

cycles won't do the job, and even if there is a major explosive volcanic eruption in the next few decades, the cooling will be short-lived.

And some climate-change skeptics think that Milanković cycles are responsible for the current climate warming. This doesn't cut any ice either, because as shown on figures 5.6, 5.8, and 5.10, we are currently in a period of decreasing 65° north summertime insolation, and that contributes to cooling, not warming.

6

MOVING HEAT
WITH OCEAN CURRENTS

*A current such that, although they had great wind, they
could not proceed forward, but backward and it seems
that they were proceeding well; at the end it was known
that the current was more powerful than the wind.*

— Observations of the Castilian explorer
Juan Ponce de Leon on the east coast of Florida in 1513[1]

JUAN PONCE DE LEON was the first to describe the Gulf Stream, although it is unlikely that he realized the significance of his obser-
vation, or the importance of that current or any other current, to the
Earth's climate.

Ocean Currents

In fact, ocean currents in general have huge implications for the
Earth's climate as a whole. The major surface currents of the world's
oceans are shown on figure 6.1. Because of the Coriolis effect, currents tend to have a clockwise pattern in the northern hemisphere
and a counter-clockwise pattern in the southern hemisphere. Currents that flow toward the equator generally bring cold water into
warmer regions (double arrows), while those that flow toward the
poles bring warm water into colder regions (dashed arrows). In most
cases, currents that flow generally east–west have a neutral role when
it comes to redistributing heat.

This redistribution of cold and warm water by currents is critical to the maintenance of a moderate climate in various parts of the Earth. Without these currents, the tropics would be uncomfortably hot and polar regions unbearably cold.

The volume of water moved by ocean currents is enormous. Where the Gulf Stream flows past Newfoundland, its flow volume is estimated to be 150 million cubic meters per second.[2] The combined flow volume of all of the Earth's rivers where they enter the oceans is approximately 1.3 million cubic meters per second, or less than 1% of the flow of just that one current.[3]

The currents shown on figure 6.1 are surface currents, and as such, they are confined to the upper 400 m of the oceans, and in fact, most of the flow is within the upper 100 m. But there are also significant deep-flow currents, and those play an equally important role in the redistribution of heat on the planet. Some of that deeper flow is shown on figure 6.2. This is known as the thermohaline circulation system, where the term "thermohaline" implies that these flows are driven, at least in part, by both the temperature and the salinity of the water.

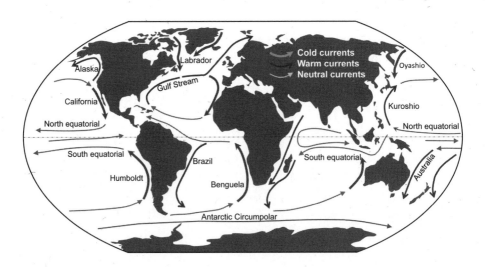

Figure 6.1. Major surface currents in the world's oceans. The "cold currents" bring cold water into relatively warm regions, and the "warm currents" bring warm water into relatively cold regions. Based on U.S. Government public domain images at "Corrientes-oceanicas.png," *Wikimedia*.

These factors determine the density of the water in the currents, and that is critical to the thermohaline circulation.

Pure water at 20°C has a density of 998 grams/liter (g/L). Typical salty ocean water (3.5% salinity) at 20°C has a density of 1,025 g/L. Ocean water salinity ranges from about 3.3% in areas where there is a lot of rain or a lot of freshwater input from rivers to about 3.8% where there is strong evaporation and relatively little freshwater input. The higher the salinity, the greater the density. Ocean water temperature ranges from around 30°C in the tropics to a little less than 0°C[4] in polar regions. The lower the temperature, the greater the density because a kilogram of cold water occupies less volume than a kilogram of warm water.

The water of the Gulf Stream flowing past Florida has a salinity of around 3.6% and a typical temperature of about 28°C. As this water flows north beyond Iceland, its salinity drops only a little, to about 3.5% (due to rain and river input), while its temperature drops a lot, to about 2°C. That cold salty water has a density of about 1,028 g/L, making it the densest water anywhere in the open ocean, and quite a lot

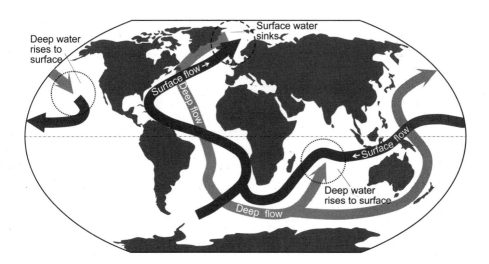

Figure 6.2. Major thermohaline flow patterns in the world's oceans. Cold salty surface water sinks in the north Atlantic, and deep water comes to surface in the Indian and the north Pacific oceans. Based on maps of thermohaline circulation from various sources.

denser than the equally cold but much less salty water underneath. Because of this high density, the surface water in that region sinks and becomes part of the deep flow system. It remains submerged as it moves south through the Atlantic and east past Africa, then it resurfaces either in the Indian Ocean—east of Madagascar—or in the northern part of the Pacific Ocean—north of Hawaii (figure 6.2). This thermohaline circulation is important in moderating the Earth's climate, and it exerts significant control over the surface currents.

Ocean Current Variations During Glaciations

Glacial ice cores collected from Greenland and Antarctica have revealed some striking temperature variations over the past few hundred thousand years, especially during the more intense parts of the Quaternary glaciations, and it has been shown that much of that variability is related to changes in current flow patterns. Figure 6.3 shows the temperature record for the past 100,000 years at the ice

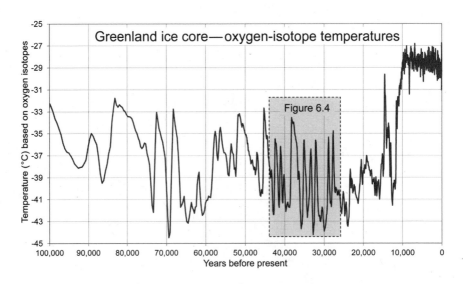

Figure 6.3. Oxygen-isotope temperatures at the GISP2 site for the past 100,000 years. Based on data at ncdc.noaa.gov/data-access/paleoclimatology -data/datasets/ice-core, as described in Alley, R., "The Younger Dryas Cold Interval as Viewed from Central Greenland," *Quaternary Science Reviews*, V. 19, pp. 213–26, 2000. The data for the shaded area are shown in more detail on figure 6.4.

surface, as determined in core samples from the GISP2 drill hole in central Greenland. This time period covers virtually all of the last glacial cycle and also the return to non-glacial conditions of the past 10,000 years.

During the last glaciation, there were repetitive temperature swings, in the order of 6° to 10°C, on time scales of 1,000 to 2,000 years. These oscillations are known as Dansgaard-Oeschger cycles in honor of the Danish (Willi Dansgaard) and Swiss (Hans Oeschger) scientists that first described them.

It is important to note that these temperatures represent the surface of the Greenland ice sheet at a site that is currently 3,200 meters above sea level. At that location, the present-day mean annual temperature is close to –25°C, and the range in mean monthly temperatures between winter and summer is about 30°C. For comparison, the global range between winter and summer is about 11°C.[5]

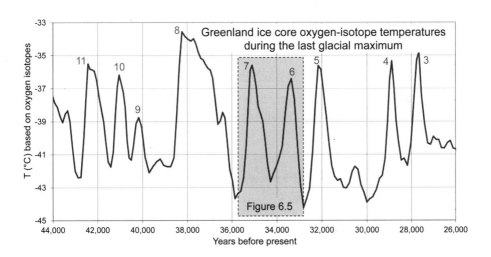

Figure 6.4. Oxygen-isotope temperatures at the GISP2 site for the period from 44 to 26 thousand years ago. Source: See Figure 6.3. The numbered peaks are known as the Dansgaard-Oeschger cycles, described in Dansgaard, W., "North Atlantic Climate Oscillations Revealed by Deep Greenland Ice Cores," in *Climate Processes and Climate Sensitivity*, J. E. Hanson & T. Takahashi, eds., Washington, DC: American Geophysical Union, pp. 288–98, 1984; and Dansgaard, W., et al., "Evidence for General Instability of Past Climate from a 250-kyr Ice-Core Record," *Nature*, V. 364, pp. 218–20, 1993.

A more detailed view of the temperature record for the latter part of the last glaciation is provided on figure 6.4. Although the timing of the peaks during this period is not entirely regular, there is a periodicity to it, and the average interval between peaks is about 1,500 years.

The origin of the Dansgaard-Oeschger cycles is not fully understood, but over the past two decades, there has been an increasing consensus, among marine scientists and climatologists, that the key factor is variability in the salinity of the northern Atlantic Ocean. This variability is known as the "salinity oscillator." As described above, evaporation in the equatorial part of the Atlantic increases the salinity of the Gulf Stream. In the far north Atlantic, this still salty, and now cold water sinks to become part of the subsurface flow that eventually comes back to surface in the Indian and Pacific Oceans (figure 6.2). That represents a net movement of salt out of the Atlantic basin, which gradually (over hundreds of years) results in a decrease in the overall salinity of the Atlantic water. Another part of the process is related to the northward transportation of heat by the Gulf Stream, which makes the Arctic region warmer than it would be otherwise, and that leads to more melting of glacial ice in Greenland and northern Canada, further diluting the Atlantic.

As the Atlantic slowly becomes less salty and as the melting of northern glaciers make it fresher still, the tendency for the cooled Gulf Stream water to sink in the far north Atlantic is reduced, and therefore the strength of the overall thermohaline circulation system decreases. That means that less heat is transported north; the Arctic region cools; less salt is removed from the Atlantic basin; and glacier-melting slows so that less freshwater flows into the ocean.

The salinity oscillator process is illustrated on figure 6.5, which is focused on Dansgaard-Oeschger events 6 and 7. The dashed gray curve is not based on data; it is simply a representation of slow changes in the salinity of the Atlantic Ocean and the strength of the thermohaline circulation (THC). When the THC is strong—because the salinity is high—western Europe and the Arctic region are warmed. A strong THC tends to slowly reduce Atlantic salinity both because salty water is moved out of the Atlantic basin and because Arctic melting is enhanced. This slowly weakens the THC, and the Arctic cools. With a weaker THC and less Arctic melting, the salinity slowly increases. The entire cycle takes about 1,500 years on average, although in the case illustrated, the time between peaks is 1,750 years.

As Figure 6.3 shows, there is no evidence of any north Atlantic salinity-oscillation temperature swings over the past 12,000 years. That doesn't mean that the salinity of the Atlantic hasn't changed during that time, only that there is no temperature record of it in ice cores. In fact, we don't know how, or if, the salinity oscillation process functions during warm periods.

There is direct evidence, however, that the Atlantic THC does change under warm conditions. This is illustrated on figure 6.6, which, based on an 1,100-year record of Atlantic Ocean temperatures, implies a weakening of the THC over the past 170 years. This weakening is also supported by changes in Atlantic salinity and by direct

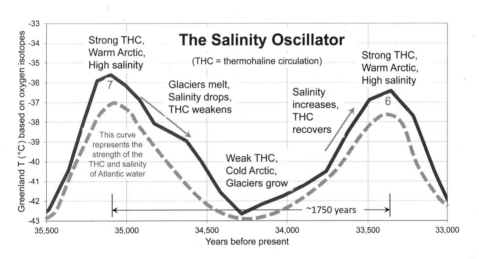

Figure 6.5. Illustration of the Atlantic salinity oscillator, showing changes in Greenland ice temperatures and the strength of thermohaline circulation (THC) and salinity of the Atlantic during Dansgaard-Oeschger events 6 and 7. Source: See Figure 6.3. The term "Atlantic THC" refers here to the overall strength of surface and deep currents in the Atlantic Basin. This is correlative with the more widely used term "Atlantic Meridional Overturning Circulation" (AMOC).

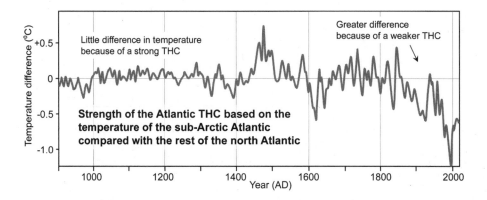

Figure 6.6. A record of changes in the strength of the Atlantic thermohaline circulation based on the record of differences in temperature in different parts of the north Atlantic. Based on Rahmstorf, S., et al., "Exceptional Twentieth-century Slowdown in Atlantic Ocean Overturning Circulation," *Nature Climate Change*, V. 5(5), pp. 475–80, 2015. Temperatures prior to 1900 are based on proxy data from marine sediment cores.

measurements of current strengths in the Atlantic basin.[6] It is not known if these changes are related to the natural salinity oscillation process described above (or to some other natural process) or to anthropogenic climate change, which has undoubtedly led to enhanced glacial melting on Greenland and in other parts of the Arctic over the past several decades.

Atlantic Thermohaline Circulation and Future Climate

The record of past millennial-scale changes in ocean currents in the Atlantic raises some interesting questions about our climate. It is evident (as shown on figures 6.4 and 6.5) that there were dramatic natural changes in temperature during the last glacial cycle. It would be interesting to know if some of these were every bit as fast as the current human-caused changes. We might also ask if the observed weakening of the THC is going to lead to significant future cooling that could help to offset human-caused warming.

Looking first at the past rate of change, we can see (from figure 6.5) that over the 350 years from 35,500 to 35,150 years ago, there was 7°C of warming, or an average of 2°C per century. But this was in the middle of Greenland at an elevation of 3,200 m, where the typical summer–winter temperature range is almost three times the global average, and where present-day climate warming is also about three times the global average. We don't know how that change on Greenland might relate to the global average change, but it's likely to be equivalent to around 1°C of *global* warming per century. The average global temperature has increased by 1°C in the last 50 years[7] (or 2°C/century), which is about twice the fastest natural rate during the last glaciation. Moreover, the warming that took place on Greenland during Dansgaard-Oeschger cycles always happened during periods of increasing THC strength. The THC has been weakening for the past 150 years, so there is no chance that this warming has been caused by that process.

And that brings us to the question of whether the observed weakening of the THC could lead to global cooling that might offset human-caused climate warming. As already noted, the climate changes related to THC oscillations took place during the height of the last glaciation. There is no evidence to suggest that those types of changes will have similar cooling results in a world that is relatively unglaciated. Furthermore, the THC has now been steadily weakening for 150 years. During most of that time, the Earth's climate has been warming, not cooling. But even if a change in the THC leads to cooling, it is likely to have only a regional effect, specifically in western Europe and the north Atlantic. In fact, during the Dansgaard-Oeschger cycling process, the observed cooling in the north Atlantic always coincided with warming in the south Atlantic, and vice versa.[8]

El Niño Southern Oscillation

Many North Americans are familiar with the El Niño phenomenon (aka El Niño Southern Oscillation or ENSO) because it has direct implications for our weather on a year-to-year basis.

As seen on figure 6.1, the Humboldt Current (aka Peru Current) brings cold water from the Southern Ocean north along the west coast of South America. Much of this water continues west across the

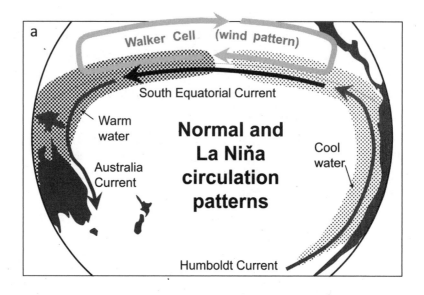

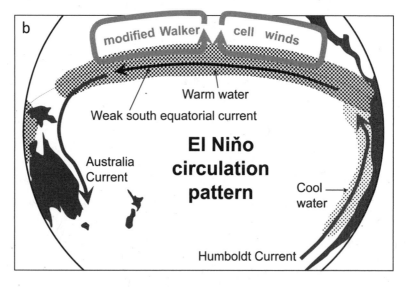

Figure 6.7. Wind and ocean current patterns in the southern Pacific Ocean. Under the normal and La Niña patterns (*a*), strong west-flowing Walker Cell winds push cool Humboldt Current water east across the Pacific, keeping the waters around Australia and Southeast Asia warm. Under the El Niño pattern (*b*), the Walker Cell has broken down, the currents have slowed, and warm water has moved east across the Pacific as far as South America. Based on various sources of information, including those used for figure 6.1.

Pacific, via the South Equatorial Current. There is a specific air circulation pattern in the equatorial Pacific called the Walker Cell that consistently moves low-level air to the west across the surface of the Pacific, up in the area around Australia and Indonesia, back to the east at higher elevation, and then down in the area around Ecuador (figure 6.7). This wind keeps warm water pushed across to the western side of the Pacific, near to Australia and Southeast Asia, allowing cold water of the Humboldt Current to keep moving north and also allowing deep cold Pacific Ocean water to continue coming to surface in that area. The Walker Cell periodically weakens, on a not very consistent cycle, allowing warm water to flow back toward South America and slowing the Humboldt Current.

The measurement of ENSO is based on the relative surface-water temperature of an area called "Niño 3.4," which is a patch of the central Pacific Ocean that straddles the equator and is half-way between South America and Southeast Asia.[9] The Niño 3.4 index for the past 50 years is shown on figure 6.8. The standard threshold for declaring an El Niño is a Niño 3.4 index above 0.5 for 5 consecutive months.

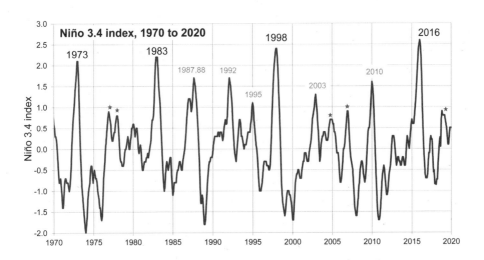

Figure 6.8. Niño 3.4 index values from 1970 to 2020. Important El Niño events are labeled. Using data from the Climate Prediction Center of the U.S. National Oceanographic and Atmospheric Administration.

On that diagram, all peaks that just meet that threshold are marked with stars, all that are above 1.0 are labelled with small year numbers, and the four that are greater than 2.0 are labelled with large year numbers.

ENSO events occur on an irregular repeat pattern of between 2 and 6 years, with an average of 3.5 years. That means that if this is an El Niño year, there could be another El Niño within 2 years, but it is more likely to be 3 to 4 years and may not happen for 6 years. Although we understand the mechanisms for changes in the ENSO patterns, we do not understand the causes of those changes.

Why Is ENSO Important?

ENSO is important because it would be useful to know whether the intensity or frequency of ENSO events has been affected by anthropogenic climate change, and whether ENSO has any implications for climate change.

There is evidence that the intensity of ENSO events is increasing. As can be seen on figure 6.8, the last four major El Niños (in 1973, 1983, 1998, and 2016) had Niño 3.4 indices of 2.1, 2.3, 2.4, and 2.6 respectively, and prior to 1970, there was no El Niño event with an index greater than 2.0. There is also a similar trend in the index values for the La Niña minimum values (they are getting warmer). While this appears to be a consistent trend, it is possible that it is only a reflection of increasing ocean temperatures. Remember that the ENSO 3.4 indices are just temperature averages, and since global sea-surface temperature has risen by a similar amount (0.5°C) over the same time period, it is quite possible that these higher index values are just a result of warmer global temperatures.[10]

There is no clear evidence from figure 6.8 that the frequency of ENSO events has changed in the past 50 years, but there is evidence from two recent studies that the area and extent of the Pacific that sees the highest water temperature is changing. In one study, core samples from corals at 27 sites across the Pacific were used to determine sea-water temperatures going back 400 years.[11] The conclusion is that, since the latter part of the twentieth century, there has been an increase in El Niño events focused in the central Pacific and a decrease in events in the eastern Pacific (adjacent to South America). The other study is based on historical data going back to 1900. The overall conclusion is similar, that the focus of El Niños has moved from the eastern Pacific to the central Pacific over the past 40 years, but also that the recent very strong El Niños (the ones with Niño 3.4 indices over 2 in figure 5.8) have had basin-wide effects.[12] These authors suggest that El Niños will continue to become more extreme in the coming decades.

Does ENSO affect our climate? The answer is yes and no. The annual average global air temperatures for the past 50 years are shown on figure 6.9, with the bars shaded depending on whether they were El Niño, La Niña, or "normal" years. It is quite clear that almost all strong El Niño years have higher-than-average global temperatures than normal years, and almost all strong La Niña years have lower-than-average global temperatures than normal years. The main reason for this is that the localized ENSO warming or cooling effects

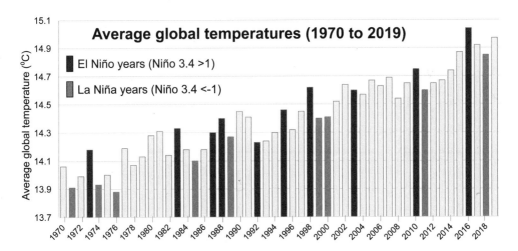

Figure 6.9. Average global temperatures from 1970 to 2019, labelled according to their ENSO status. Using data for land and sea areas from the GISTEMP Team: GISS Surface Temperature Analysis (GISTEMP), version 4, NASA Goddard Institute for Space Studies, 2020. Dataset at data.giss.nasa.gov/gistemp, accessed April 4, 2020; Lenssen, N., et al., "Improvements in the GISTEMP Uncertainty Model," *J. Geophys. Res. Atmos.*, V. 124, pp. 6307–26, 2019.

are so strong that they significantly affect the average global temperature. In some El Niño years, the global temperatures are in the order of 0.2°C higher than the years on either side.

This is important because climate-change skeptics have often seized on the fact that the climate doesn't get warmer every year, and that sometimes there have been stretches of several years with relatively stable temperatures. The best example of this was in the period following 1998, a strong El Niño year that was very warm. The following two years were much cooler, and after that the temperature trend seemed to remain almost flat until 2004. In the several years following 1998, climate-change sceptics couldn't stop talking about how the temperatures weren't increasing anymore; however, that argument started to grow weak by about 2005, and there wasn't much talk of it after 2010.

For the past 120 years, the global temperature trend has consistently been generally upward. ENSO does affect the climate on a time scale of a few years, making the climate record look spiky, but there is no evidence that ENSO affects the Earth's climate over the longer term—decades to centuries.

SHORT-TERM SOLAR VARIATIONS

The sun was orange in color. Within it there was a black vapor like a flying magpie. After several months it dispersed.

> — 5th year of the Chung-p'ing reign, AD 188

Within the sun, a three-legged crow was seen.

> — 3rd year of the Ch'ih-wu reign, AD 240

Within the sun there was the form of a flying swallow. After several days/months it dispersed.

> — 9th year of the Yuan-k'ang reign, AD 299[1]

THE CHINESE WERE CAREFUL observers of many different types of celestial phenomena, and they were likely the first to observe sunspots—long before telescopes—at around 800 BC, although the records are sporadic. The first systematic record of sunspots is credited to Johannes Fabricus, who observed the sun over several months in 1611 from his father's church in Osteel, Germany, using an early telescope brought from the Netherlands.[2] Two years later, in Florence, Galileo Galilei observed sunspots, using a telescope of his own making. He published a series of drawings made daily over 36 days, allowing him to demonstrate the changing forms of sunspots over time.[3]

What Is a Sunspot?

A sunspot is an area on the surface of the sun that appears dark because it is cooler than the surrounding area. The temperature of a sunspot is about 4,000°C, or about 1,500° cooler than the average temperature of the sun's surface. They range in size from about one-hundredth to ten times the diameter of the Earth, with the average being close to the Earth's diameter (12,000 km), or roughly 1/100 of the sun's diameter.

Sunspots result from perturbations of the sun's magnetic field and form where loops of the field penetrate the sun's surface (photosphere), as shown on figure 7.1. The penetration of the field through the photosphere inhibits the normal convective motion within the sun's upper layer, resulting in less heat getting through to the surface. Sunspots typically form in pairs, and because one of each pair is the result of the magnetic field flowing out of the surface, and the other is the result of it flowing into the surface, the two related spots have opposite magnetic polarities. Sunspots do move around a little on the sun's surface, especially when they first form. The larger ones tend to move more than the smaller ones.[4]

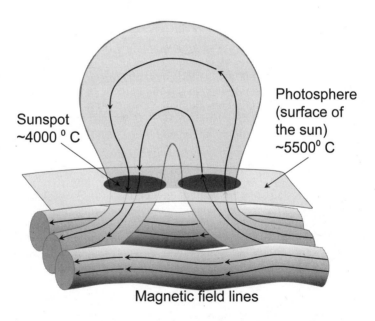

Figure 7.1. Depiction of the relationship between solar magnetic field perturbations and sunspots. Based on a drawing by Luis María Benítez, "Sunspot Diagram.svg," *Wikimedia*.

Some sunspots are shown on figure 7.2. The dark part of a sunspot is called the umbra, and the surrounding less dark halo is the penumbra. Because sunspots have lower temperatures than the surrounding regions of the photosphere, they result in a reduction of the amount of solar irradiance received on Earth. But sunspots are associated with faculae, bright areas of the photosphere that typically surround sunspot regions. Faculae form where the solar magnetic field comes near to but does not penetrate the photosphere. In these regions, the magnetic field reduces the density of gases above the photosphere,

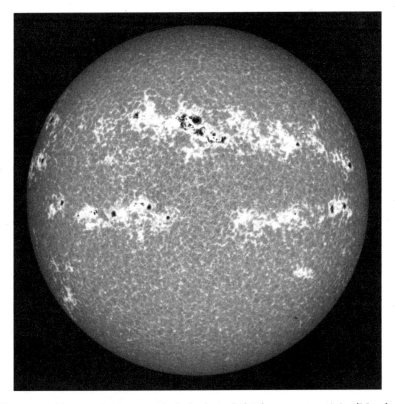

Figure 7.2. The sun during a period of relatively high sunspot activity (March 28, 2001). The sunspots are black; the gray areas around them are their penumbras; and the white areas are faculae. Image from the Precision Solar Photometric Telescope (PSPT) at the Mauna Loa Solar Observatory, processed by the NASA/Goddard Space Flight Center Scientific Visualization Studio, svs.gsfc.nasa.gov/2644.

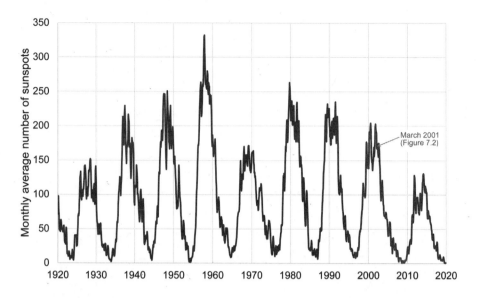

Figure 7.3. Variation in monthly average number of sunspots for the period from 1920 to 2020. Using data from SILSO, World Data Center, Sunspot Number and Long-term Solar Observations, Royal Observatory of Belgium, on-line Sunspot Number catalogue, sidc.be/silso/datafiles.

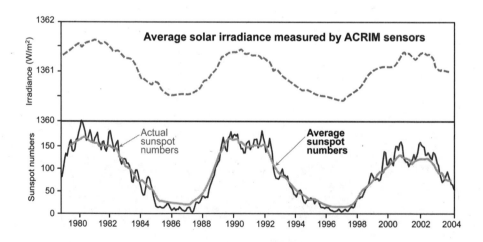

Figure 7.4. Comparison between solar irradiance (measured using satellites) and sunspot numbers over the period 1979 to 2004. Based on a NASA figure, spacemath.gsfc.nasa.gov/weekly/Earth8.pdf. Active Cavity Radiometer Irradiance Monitor sensors were used to measure solar irradiance on three different satellites between 1980 and 2013, at altitudes of 500 to 700 km.

allowing light to shine through with greater intensity. Because of the larger extent of the faculae (the white areas on figure 7.2) compared with the area of sunspots, the overall solar irradiance is actually greater when sunspot numbers are high than when they are low.

Short-term Sunspot Variations

The number of sunspots visible at any moment ranges from none to several hundred. There is a regular pattern of variation in sunspot numbers, with a period close to 11 years, although there are other longer-term variations that do not appear to be regular. The 11-year sunspot periodicity has recently been correlated with an 11-year cycle of the three planets that have the greatest gravitation influence on the sun: Venus, Earth, and Jupiter.[5] This work has shown that the tidal pull of Venus, Earth, and Jupiter is strongest when these planets line up on the same side of the sun every 11.07 years, and that these alignments consistently correspond with the sunspot minima.

Monthly sunspot values for the past century are shown on figure 7.3. The 11-year cycle is clearly evident, and it's also evident that there is significant month-to-month variability. The period around March 2001 is identified (with approximately 160 sunspots), as that is what is shown on figure 7.2. (Yes, it's true that the figure does not show 160 sunspots but many of the smaller sunspots are not visible, and the other half of the sun is also not visible.)

Comprehensive records of sunspot numbers date from about 1750, but there are enough telescope observations to piece together a reliable sunspot history going back to 1611.[6] The naked-eye observations prior to 1611 are too sporadic to allow extending the record any further back than that.

The relationship between sunspot numbers and solar irradiance is shown on figure 7.4. The correlation between the two parameters is very clear, but the variation in irradiance is small. When the average number of sunspots is around 150, the average irradiance is about 1361.5 Watts per square meter (W/m^2).[7] When the average number of sunspots is close to zero, the average irradiance is about 1360.5 W/m^2. In other words, there is a difference of one W/m^2, or less than 0.1%, in the amount of energy we get from the sun during a relatively high sunspot period compared with a low sunspot period. If you are thinking that this cannot possibly make any difference to the climate, you may be right, and you may be wrong.

You may be right because, on the 11-year time scale of the sunspot cycles, the small differences in solar irradiance do not change the climate because our climate system needs time (decades) to respond to external forcings. In the absence of climate feedbacks, a 0.1% change in irradiance should lead to a 0.03°C change in the atmospheric temperature at surface once equilibrium has been reached. Because of the mass of the oceans and the land, equilibrium is not even approached for decades, and not reached for centuries. So, long before our climate starts to come into equilibrium with a change in sunspot numbers, the cycle is over, and the numbers have changed again.

On the other hand, you may be wrong because of feedbacks but also because there are some longer-term variations in sunspot numbers. The tiny temperature forcing related to sunspot variations can be amplified many times by climate feedbacks, and it is possible that longer-term variations—but not the short ones—could change our climate.

Long-term Sunspot Variations and the Little Ice Age

There are two types of long-term changes in sunspot cycles. Firstly, there are the regular variations in the intensity of sunspot-number peaks, as shown on figure 7.3. Sunspot numbers were relatively low in the 1920s, with a maximum of about 150 sunspots. The numbers increased through the first half of the twentieth century to a maximum of over 300 sunspots in the late 1950s, and then gradually dropped again to a low maximum of about 130 sunspots in the 2010s. As shown on figure 7.5, a similar long-term trend existed from about 1820 to 1900. The period of low sunspot numbers at around 1800 is called the Dalton minimum, after English meteorologist John Dalton. There is no consistent evidence that the Dalton sunspot minimum, which lasted about 30 years, had a measurable impact on the climate.[8]

Secondly, there have been periods of many decades with very few sunspots. These are called "grand minima," and the best known of them is the Maunder minimum, the only one for which we have telescope observational data. There were fewer sunspots observed during the entire 70-year period from 1645 to 1715 than during any single year in the twentieth century. Most of those were single spots that were short-lived, lasting for less than a single solar rotation (26 days as observed from Earth).[9]

The irradiance values, shown as a heavy black line on figure 7.5, are based on beryllium-10 (^{10}Be) concentrations in glacial ice and carbon-14 (^{14}C) in tree rings, both from multiple sources. These isotopes of beryllium and carbon are naturally formed in the upper atmosphere as a result of the interaction of cosmic radiation with atmospheric nitrogen and oxygen. The sun's magnetic field, which increases in intensity with the number of sunspots, deflects cosmic radiation away from the Earth, and so variations in the concentrations of ^{10}Be and ^{14}C on Earth are proportional to the number of sunspots.

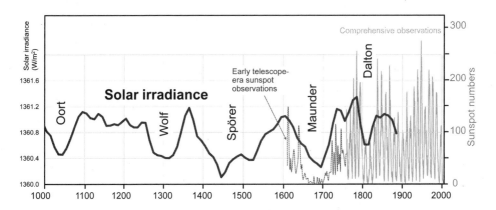

Figure 7.5. Sunspot numbers from 1750 to 2020 (in gray) and solar irradiance from 1000 to 1885 (in black) based on tree-ring carbon-14 and ice-core beryllium-10 values. Using sunspot data from SILSO, World Data Center (same source as figure 7.3), and irradiance data from Wu, C., et al., "Solar Total and Spectral Irradiance Reconstruction over the Last 9000 Years," *Astronomy and Astrophysics*, V. 620, A120, 12, 2018; www2.mps.mpg.de/projects/sun-climate/data.html.

Although there were no consistent sunspot observations prior to 1611, it appears from the irradiance data that the Spörer minimum was just as deep as or even deeper than the Maunder minimum, and so we can assume that there were long periods between 1400 and 1550 with few or no sunspots. The Wolf and Oort minima were less extreme, although it is also likely that there were lengthy periods without sunspots during the Wolf minimum (1280 to 1350).[10]

The solar irradiance lows of the various grand minima are shown on figure 7.5. Each lasted for several decades (as compared with just a few years for normal minima), and so there is a possibility that they had a climate impact. In fact, there is a coincidence between the period that includes the Wolf, Spörer, Maunder, and Dalton minima and that of the Little Ice Age, as shown on figure 7.6.

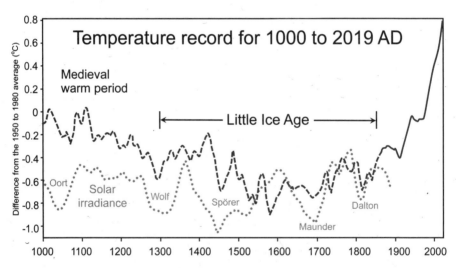

Figure 7.6. Global temperature record for the period AD 1000 to 2019, from proxy sources (dashed line) and from actual readings (solid line). The solar irradiance curve (from figure 7.5) is included for reference. Based on Moberg, A., et al., "Highly Variable Northern Hemisphere Temperatures Reconstructed from Low- and High-Resolution Proxy Data, *Nature*, V. 433, pp. 613–17, 2005 (pre-1900); and Jones, P., and Moberg, A., "Hemispheric and Large-scale Surface Air Temperature Variations: An Extensive Revision and an Update to 2001," *Journal of Climate*, V. 16, pp. 206–23, 2003.

The Little Ice Age (LIA) was a period of consistently cool weather (and some particularly cold winters) that lasted for over 500 years from around 1300 to 1850,[11] although the timing varies depending on the location and the type of evidence available for interpretation. It wasn't a real ice age—no continental glaciers formed—and it wasn't cold for that entire time, although there were some particularly cold periods, such as from 1450 to 1475 and from 1645 to 1715. During the Little Ice Age, glaciers advanced by hundreds to thousands of meters, in some cases crushing alpine villages, such as La Rosière in France (1616), Pre du Bar in Italy (1715), and a Tlingit village in Alaska. Significant glacial advances were also seen in central Asia.[12] In fact, glaciers throughout the northern hemisphere advanced significantly during the LIA.

The Robson Glacier in the Canadian Rocky Mountains provides an example of a glacial advance around the time of the Little Ice Age (figure 7.7). At the height of the last glaciation, about 20 ka (20,000 years ago), the entire valley shown on figure 7.7 was filled with ice and the glacier extended nearly 20 km down the valley, only to join an even larger glacier there. Those glaciers all started receding at around 12 ka because of warming related to increased Milanković insolation, and by about 5.5 ka, this glacier had receded to a point about 4 km above its current terminus (behind Rearguard Mountain in the photo).[13] At that time, it started to readvance, and it reached the limit shown by the terminal moraine at point B on figure 7.7 by around AD 1350. There is no evidence that this glacier advanced beyond that point during the rest of the LIA.[14]

A similar history is recorded for Switzerland's Great Aletsch Glacier in (the longest glacier in Europe), as illustrated in figure 7.8. From around 3,500 years before the present until approximately AD 1350, the glacier's terminus advanced and retreated several times, but its overall advance was in the order of 3 km. During the LIA, it also retreated and advanced, but it never advanced more than it had by AD 1350, and it never retreated more than approximately 1.5 km. Since 1850 it has retreated approximately 3.5 km.

Both of these glaciers expanded significantly over the past several thousand years, reaching maximum extents during the LIA, but these glacial expansions, which began millennia ago, cannot be correlated with the sunspot grand minima that took place between 1050 and 1800. Instead, they appear to be more closely correlated with a drop in Milanković insolation at 65° N that has taken place over the past several thousand years (figure 7.8) but actually started about 10,000 years ago. This conclusion is supported by Solomina and others[15] who reviewed data from more than one hundred glaciers on all of the continents (except Australia) and noted a general trend of increasing glacier size in northern hemisphere glaciers during the past 10,000 years, corresponding with a decline in summer-time insolation at 65° N.

Figure 7.7. Rearguard Mountain (right) and the Robson Glacier, Rocky Mountains, British Columbia, 2018. The trim line along the far slope (A) marks the maximum height of the Robson Glacier within the past 5,000 years. At that time (AD 1350), the terminus was at the location of the crescent-shaped terminal moraine (B). The present terminus is now 2.2 km up the valley (C).

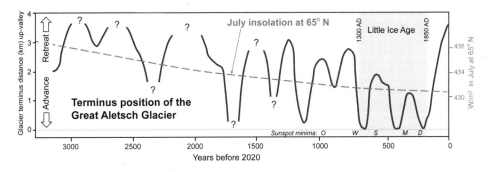

Figure 7.8. Variations in the position of the terminus of the Grosser Aletsch Glacier (Switzerland) relative to its maximum extent during the LIA. The sunspot minima are shown with the letters O (Oört), W (Wolf), S (Sporer), M (Maunder), and D (Dalton). From Matthews, J., and Briffa, K., "The 'Little Ice Age': Re-evaluation of an Evolving Concept," *Geografiska Annaler*, V. 87, pp. 17–36, 2005. The insolation curve is from data available at vo.imcce.fr /insola/earth/online/earth/earth.html; Laskar, J., et al., "A Long-Term Numerical Solution for the Insolation Quantities of the Earth," *Astronomy and Astrophysics*, V. 428, pp. 261–85, 2004.

In other words, the evidence suggests that the Little Ice Age was more likely to have been caused by insolation changes related to orbital variations than by low sunspot numbers, although it is still quite possible that the cold conditions were exacerbated by a small reduction in solar irradiance.

Sunspots and Recent Climate Change

One of the common arguments used by those who don't think that humans are causing the strong warming that we've seen over the past century is that "it's the sun." They might be referring to changes in insolation related to the changing orbital cycles, or they might be referring to changes in solar irradiance related to sunspot cycles. Either way, there is no evidence to support these claims.

As discussed in chapter 5, and shown on figure 7.8, the Milanković cycle change in July insolation at 65° N over the past several thousand years can only have led to planetary cooling, not warming. This appears to have been what brought the Earth into the Little Ice Age around 700 years ago, but that cooling effect hasn't reversed in the past century, so it cannot be used to explain our current warming climate.

The evidence that sunspot variations have significantly affected out climate in the past is weak at best, but there is no evidence that this is happening now. As shown on figure 7.9, total solar irradiance (related to sunspot cycles) increased by about 1 W/m² from 1880 to 1950. There was about 0.2°C of warming over that time, but it is unlikely that it is related to sunspots and solar irradiance. Since 1955 the solar irradiance has dropped by more than 0.5 W/m². During that time, there has been more than one degree of warming.

The changes both in the amount of solar radiation emitted (due to sunspot changes) and in the amount received in the Earth's northern hemisphere (due to orbital variations) should have led to a small amount of cooling over the past 70 years, not to the strong warming that we have observed.

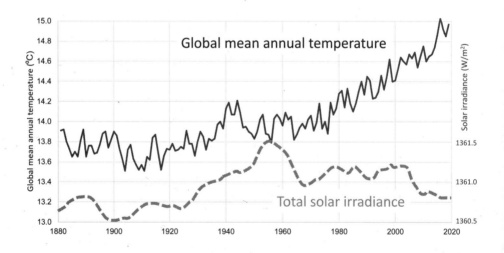

Figure 7.9. Global mean annual temperatures over the past 140 years versus total solar irradiance related to sunspots. Based on temperature data from NASA Goddard Institute for Space Studies; and on solar irradiance data from Dasi-Espuig, M., et al., "Reconstruction of Spectral Solar Irradiance Since 1700 from Simulated Magnetograms," *Astronomy & Astrophysics*, V. 590, 2016.

8

CATASTROPHIC COLLISIONS

*An airliner flies at an altitude of about 10 km, so imagine
a plane unfortunate enough to be in the way of the
incoming comet. In an instant the airplane would be
smashed like a bug by the onrushing body. One-third
of a second later the front of the comet, carrying the
insignificant aircraft wreckage, would hit the ground,
generating a blinding flash of light and initiating shock
waves in the comet and the ground, and after another
⅓ second the back end would be passing below ground
level. By one or two seconds after the loss of the airplane,
there would be a huge, growing, incandescent hole in
the ground and an expanding fireball of vaporized rock,
and debris ejected by the explosion would be clearing the
atmosphere on its way to points around the globe. Earth
would suffer cataclysmic damage in less time than it
takes to read this sentence.*

— Walter Alvarez, 2013[1]

I T TOOK ME ABOUT FIVE SECONDS to read that last sentence, and
for almost all organisms living within a few hundred kilometers of
the impact site, close to the pueblo of Chicxulub, Yucatan, Mexico,
that would have been all the time they had. Further away—up to at
least 6,000 km but likely more—death for most organisms would
have come within hours in the form of radiative heat, not from the

impact blast itself, but from intense light radiating from uncountable incoming fragments of incandescent glass and rock that had been blasted out of the crater and out of the atmosphere, but then returned a few hours later. The luminescence of that intense "meteor" storm is estimated to have been about seven times that of the sun at its brightest, or roughly equivalent to the heat you would experience inside your kitchen oven on "broil." It lasted for several hours.[2] Except in areas with thick cloud cover—although even that would likely have evaporated—most flammable material, forests for example, would have ignited, and most exposed organisms would literally have been broiled. A 6,000-km radius from Yucatan is enough to cover all of North America and most of South America. There's a good chance it went well beyond that.

That was just the first few hours. In the following months and years—if the animals of that time could have had such thoughts—most that survived that horror might have wished they hadn't.

Climate Effects of the Chicxulub Impact

It is estimated that the Chicxulub object was approximately 12 km in diameter and that it hit the Earth at about 100,000 km/h.[3] It landed in an area of relatively shallow sea water (a few hundred meters depth) that was underlain by 5-km of limestone and evaporite rocks. Limestone is primarily calcium carbonate ($CaCO_3$); a major component of the evaporite layer is calcium sulphate ($CaSO_4$). The impactor exploded through those sedimentary layers in a fraction of a second and then excavated a hole nearly 100 km wide and 20 km deep into the granitic rocks of the crust (figure 8.1). All of that material was immediately melted or vaporized, and the solid particles were blasted out of the atmosphere.

The sudden change in the level of the seafloor within and around the crater created an enormous tsunami that spread across the Gulf of Mexico and out into the Atlantic and to the Pacific (as this predates the Isthmus of Panama). Within the gulf, the wave would have been as high as 1,500 m, and it was likely more than 15 m high in the Pacific and Atlantic basins.[4] This wreaked massive destruction in the extensive low-lying areas around the gulf, and then the water rebounded back toward Chicxulub, refilling much of the crater with sediments from around the region and with debris from the blast. The back-filled material included a large volume of charcoal, presumably from wildfires in areas around the gulf.

As already described, the rock fragments and glass of the impact plume were sent aloft through the atmosphere and dispersed across an area with a radius of at least 6,000 km. On reentry, friction turned this material into glowing bodies with enough intensity to outshine the sun by a factor of seven times and enough heat to start wildfires within that vast area, and possibly almost everywhere on Earth. Animals that could not hide underground or in water would likely have perished in the resulting fires, if they hadn't already been killed by the incandescent heat.

Wildfires of the magnitude described above would have created a massive amount of smoke; the layer of soot from that smoke has been found in the Cretaceous-Paleogene (K-Pg) boundary-layer deposits around the world. It amounts to approximately 15 giga tonnes (Gt), which is many hundreds of times the amount of soot produced by

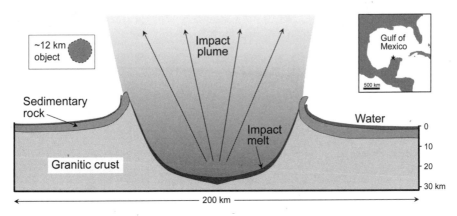

Figure 8.1. A depiction of the scenario at Chicxulub within the first minute of the impact of a 12 km diameter object. Based on a drawing in Gulick et al., "The First Day of the Cenozoic," *Proc. Natl. Acad. Sci.*, V. 116, no. 39, 2019.

wildfires in a typical year, enough to have effectively blocked most of the incoming sunlight. A digital model of the early Paleogene climate following the emission of 15 Gt of wildfire soot shows that for several years the average insolation at the Earth's surface was likely less than 1% of normal, and for low-latitude regions it was probably only a fraction of that.[5] In other words, it was permanently dark, although not quite as dark as nighttime, for years. Under those conditions, photosynthesis, and therefore growth of plants, on both land and in the oceans, was effectively stopped. Mean annual temperature in most continental areas, including equatorial regions, dropped to less than 0°C, or 10° to 15°C below normal (figure 8.2); in some temperate and polar areas, it was much less, although it probably stayed above freezing over tropical and temperate oceans. Precipitation dropped to about 20% of normal levels, typical of deserts in most regions (figure 8.2). Bardeen and co-authors conclude that the darkness would have started to lift after about two years, but the cold and dry conditions likely persisted for six to eight years.

Needless to say, survival would have been difficult under such con-

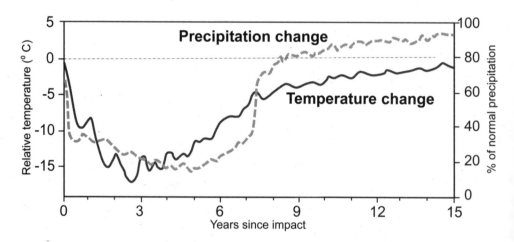

Figure 8.2. Modeled temperature (global, land, and sea) and precipitation changes in the 15 years following the K-Pg impact, based on 15 Gt of soot in the atmopshere. Based on drawings in Bardeen, C, et al., 2017, On transient climate change at the Cretaceous-Paleogene boundary due to atmospheric soot injections. Proc. Natl. Acad. Sci. V. 114, E7415-E7424

ditions, even for the fittest. Animals that had survived the heat blast and the raging wildfires would have emerged from burrows, crevices, and swamps to find permanent near darkness, no new plant growth, wicked cold, and, soon, almost no fresh water.

And that's not all; the modeling described above didn't account for the 650 Gt of sulphur dioxide that would have been formed by the instantaneous vaporization of the thick beds of calcium sulphate, roughly 100 times the amount produced by the climate-cooling eruption of Pinatubo in 1991. This SO_2 would have quickly been converted to H_2SO_4 droplets (sulphuric acid) in the atmosphere, which as long as they stayed up there (likely a few years), would have made the cooling even stronger. When significant rain finally returned, likely sometime in the sixth or seventh year, it would have been acidic.

In addition to soot, the wildfires would have produced a massive amount of CO_2, and CO_2 was also emitted by the vaporization of limestone at the impact site. Analysis of fish remains in a section spanning the K-Pg boundary in Tunisia provides evidence that, once the dust and sulphate aerosols had settled, there was approximately 5°C of CO_2-warming of the climate, and that warming lasted for about 100,000 years.[6]

To summarize, the collision of a 12-km-diameter extraterrestrial object with the Earth at 66 Ma appears to have led to both immediate and long-term climate effects. First, there was intense heat from incoming solid impact ejecta. This lasted for several hours and was enough to kill exposed organisms and start wildfires that covered entire continents. Next came several years of darkness, strong cooling, and aridity, followed by acid rainfall. Finally, there was strong warming that lasted for about 100,000 years.

About 75% of species went extinct as a result of the K-Pg event, but what is most remarkable, considering the misery that they went through, is that 25% did not. The following is a list of some of the groups that did not survive:
- more than 50% of land plant species
- 80% of turtle species
- 50% of crocodilian species

- 100% of pterosaurs
- most bird species
- 100% of dinosaurs (except for those bird species that survived)
- most metatherian mammals, including almost all marsupials
- all ammonites except nautiloids
- almost all ostracods (shrimp-like arthropods)[7]

History of Extraterrestrial Impacts

Although there are hundreds of impact craters on Earth, most are small (a few kilometers across or less), and most are relatively young (including the K-Pg impact). Details of some of the major impact craters are listed in table 8.1. The most visually striking of those craters, Manicouagan, is illustrated on figure 8.3. The main reason that we don't see a lot of older craters on Earth is that, unlike the Moon, ours is a geologically active planet. Most of the older rocks of the crust have been subducted into the mantle, or folded and faulted into mountain ranges and then eroded or covered up with hundreds of meters of younger rocks.

Table 8.1. Nine of the Earth's largest impact craters, arranged in order of date (Ma = millions of years ago; the dates are approximate because it is difficult to date impacts). Based on information from various sources.

Name	Location	Crater diameter	Date (Ma)
Vredefort	Free State, South Africa	300 km	2020
Sudbury	Ontario, Canada	260 km	1850
Acraman	South Australia, Australia	90 km	580
Manicouagan	Quebec, Canada	100 km	214
Morokweng	Northwestern South Africa	70 km	145
Kara	Nentsia, Russia	65 km	70
Chicxulub	Yucatan, Mexico	150 km	66
Popigai	Siberia, Russia	100 km	35
Chesapeake	New Jersey & Delaware, US	85 km	35

In fact, the rate of incoming objects was much higher in the Earth's early history than it is now, and this is well illustrated by an image of Mare Crisium, one of the Moon's major plains (figure 8.4). The lunar crust around Mare Crisium is around 4 Ga (4 billion years ago), while that of the floor of the mare is filled with basaltic rock that is younger, around 3.2 Ga (or possibly less). The older rock is pitted with craters of all sizes, while the mare plain is relatively pristine, with only a few small craters. The Moon was bombarded with objects for the first 500 million years of its history, during what is known as the Early Bombardment (from around 4.5 to 4.0 Ga). This was followed by the Late

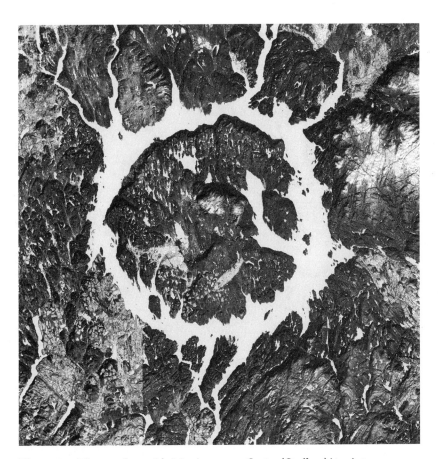

Figure 8.3. The 100-km-wide Manicouagan Crater (Québec) in winter. "Manicouagan Reservoir in winter by Sentinel-2.jpg," *Wikimedia*, May 2020.

Heavy Bombardment, which peaked at around 3.9 Ga, and gradually decreased in intensity until around 3.0 Ga.[8] The Mare Crisium flood basalts covered up a lot of older craters and have not been heavily cratered themselves because of the much lower rate of bombardment after 3 Ga.

The important point here is that the Earth was also bombarded with extraterrestrial objects during the period from 4.6 Ga to 3.0 Ga. We don't have evidence of these impacts because of the Earth's strong geological recycling, but we know they happened, and it is likely that the larger ones had some climate impacts.

Figure 8.4. The Moon's 555-km-wide Mare Crisium, which was flooded with basaltic magma around 3.2 Ga. NASA image (Lunar Reconnaissance Orbiter), "Mare Crisium (LRO).png," *Wikimedia*, accessed November 2020.

So, what were the climate impacts of these known collisions, and of the earlier ones that we don't know about? Almost certainly less—in most cases a lot less—than the K-Pg impact. Some particulars about the K-Pg event that resulted in such a significant climate impact were as follows: based on the size of the crater, it was caused by a large body; the Earth was widely forested at the time, and much of that vegetation was consumed by flames, sending soot and CO_2 into the atmosphere; the object struck an area with a thick sequence of carbonate- and sulphate-bearing rocks; and the atmosphere had low levels of greenhouse gases relative to some earlier times.

It is likely that the climate and biological effects of the Precambrian (prior to 540 Ma) impacts were minimal, mostly because there was virtually no life on land, so there was no potential for wildfires. If there was significant warming or cooling, it would not likely have had a profound effect on marine microorganisms as ocean temperatures tend to change much less than land temperatures.

Ever since the proposal of an extraterrestrial cause of the K-Pg extinction by the Alvarez team in 1980, and especially since the discovery of the Chicxulub Crater in 1991,[9] geologists have been looking for evidence that some of the other major extinctions were caused by extraterrestrial impacts. The five major extinctions of the past 540 million years are the Ordovician-Silurian (444 Ma), the late Devonian (370 Ma), the end-Permian (251 Ma), the end-Triassic (201 Ma), and, of course, the Cretaceous-Paleogene (66 Ma). So far, nobody has been able to find convincing evidence that any of these extinctions, other than the last one, is linked to the impact of an extraterrestrial body.

There was an ice age (not an extinction) in the middle of the Ordovician Period that some think is related to an extraterrestrial event but not to a specific impact. There is strong evidence that, about 468 Ma, a solar-system object, now referred to as the "L-chondrite parent body" (or LCPB) broke into billions of small pieces (likely as a result of a collision in space with another object). L-chondrite meteorites are very common on Earth (about one-third of all meteorites are of this type) and have been arriving here ever since that event. But there was a particular abundance of them in the million or so years following the

LCPB breakup, and it has been suggested that the amount of meteoritic dust in the atmosphere would have blocked enough sunlight to force the cooling that led to an intense glaciation.[10]

There is also some evidence that the impact of the Popigai object in Siberia, which has been dated at around 35 Ma, was responsible for a minor extinction event at the end of the Eocene (33.9 Ma). This interpretation is based on a reevaluation of the timing of Popigai, suggesting that its arrival might have happened closer to the end-Eocene extinction.[11]

Future Extraterrestrial Impacts

The conditions during the first decades of the Paleogene Period (following the K-Pg impact) were quite horrific. Nobody wants to go through that again, so it is important for us to know something about our risk of experiencing a dangerous extraterrestrial impact in the future.

We can get a feeling for the risk of a future large object by looking at the known rates of extraterrestrial particles that reach the Earth's atmosphere. As shown on figure 8.5, there are a lot more incoming small particles than large particles; in any single year, the number of particles less than 1 mm in diameter can be expressed in the quintillions (1 quintillion is 10^{18}), with many times more of cometary origin than of asteroid origin. Particles that size don't even become meteors, meaning that they don't create a visible glowing track across the sky, and they don't significantly affect our climate. Meteors number in the billions each year. A fireball is a particularly bright meteor (brighter than the planet Venus), and those number in the tens of thousands per year. To form a crater on Earth, an object must be well over a meter in diameter and must reach the Earth's land surface (some explode in the atmosphere, others land in water). We can expect a 5-m-diameter object (the size of a large pickup truck but the mass of a locomotive) to enter the Earth's atmosphere each year, a 10-m object (the size a locomotive but the mass of 10-car train) once every 10 years, a 20-m object once every 100 years, and a 50-m object once every 1,000 years.[12]

To put this into perspective, a 20-m object exploded over Chelyabinsk in the Ural region of Russia in 2013. Over 100 people were hospitalized, and there was more than $30 million in damage to buildings. Prior to exploding, the object was brighter than the sun, and many people experienced serious "sunburn." Just over 100 years earlier (1908), a 65-m object exploded between 6 and 10 km above the surface of the remote Tunguska region of Siberia. Trees were flattened over an area of 2,000 km². If this event had taken place in a populated region, the death and injury toll could have been in the millions and the cost to infrastructure in the billions.

Extraterrestrial objects ranging in size from meters to hundreds of meters do not change the climate, but larger ones obviously can, and

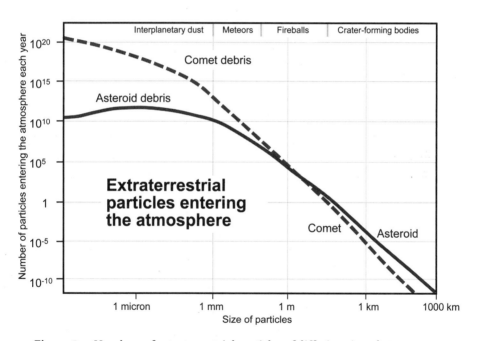

Figure 8.5. Numbers of extraterrestrial particles of differing sizes that enter the Earth's atmosphere. By the author, based on a figure in Zolensky et al., "Flux of Extraterrestrial Materials," in D. Lauretta and H. McSween, eds., *Meteorites and the Early Solar System II*, Tucson: University of Arizona Press, pp. 869–88, 2006.

figure 8.4 shows that a 1-km object can be expected about once every 100,000 years.

An international program of tracking potential "near Earth objects" (NEOs) has been in progress for the past 25 years, and over that time, thousands of objects have been identified and their orbits characterized. A potentially hazardous NEO is one that is projected to approach Earth closer than 20 times the distance between the Earth and Moon. The current focus is on all NEOs that are over 140 m in diameter, as those are considered to represent significant hazards to people and infrastructure.

According to the NASA Jet Propulsion Laboratory, there are estimated to be about 1,000 NEOs greater than 1 km in diameter and about 15,000 greater than 140 m. As of May 2020, 731 NEOs greater than 1 km and 8,827 greater than 140 m have been discovered.[13] Of these, only 2 have a significant probability of coming close to the Earth. The probabilities of those actually hitting the Earth are in the order of 1 in 10,000, and the estimated dates are more than a century away. That represents a very small risk, but it's important to remember that it is very likely that there are still hundreds of other NEOs that have not yet been discovered.

So, the risk of a large impact seems to be very small, at least in my lifetime and probably in yours. But the implications for us on Earth being hit by something several kilometers across are so extreme that it really is a good idea to keep an eye on the sky.

9

A PLAGUE OF HUMANS

*One can see from space how the human race has
changed the Earth. Nearly all of the available land has
been cleared of forest and is now used for agriculture
or urban development. The polar icecaps are shrinking
and the desert areas are increasing. At night, the Earth
is no longer dark, but large areas are lit up. All of this
is evidence that human exploitation of the planet is
reaching a critical limit. But human demands and
expectations are ever-increasing. We cannot continue to
pollute the atmosphere, poison the ocean and exhaust the
land. There isn't any more available.*

— Stephen Hawking, 2007[1]

THE AVAILABLE EVIDENCE shows that the genus *Homo* evolved at
least 2.1 million years ago.[2] For 95% of that time, our distant ancestors had relatively little impact on ecosystems and none on the climate. But as Stephen Hawking has pointed out, that all changed with
Homo sapiens, and this is especially so in the past several thousand
years.

H. *sapiens* originated in Africa sometime between 350 and 250 ka
(thousands of years ago) (figure 9.1). While still confined to Africa,
our ancestors lived through two long glacial periods and three relatively short interglacial periods. During peak glacial times, Africa
was slightly cooler and considerably drier than it is now. The Sahara

and Kalahari Deserts were more extensive, most of the current central Africa rainforest area was open-canopy savannah, while most of the current savannah areas were grassland. Surviving natural climate changes was not that difficult for *H. sapiens* in Africa, mostly because the changes came relatively slowly and there was time to adapt and to migrate.

H. sapiens started dispersing from Africa near to the end of the last interglacial, at around 120 ka, first to southern Asia, then to Southeast Asia and Australia at 70 to 65 ka, to the Middle East and Europe, and separately to Japan and Korea between 55 and 40 ka, and then to North and South America sometime between 24 and 15 ka.[3] Throughout this period, our ancestors followed a lifestyle of hunting and gathering. Their impacts were limited because they lived in small groups, they rarely threatened the viability of their food sources, and they did not destroy forests, at least not on purpose.

As we've already seen, climate variations on the time scale of tens to hundreds of thousands of years are driven primarily by the Earth's orbital and tilt cycles. This is illustrated in chapter 5, particularly on figures 5.8 and 5.9. The last 14,000 years of the ice-core climate record is depicted on figure 9.2a along with the insolation curve for

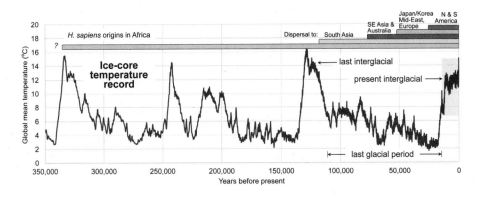

Figure 9.1. Temperature record since 350 ka (based on ice-core data) showing the origins and dispersal of *H. sapiens*. Based on data from Jouzel, J., "EPICA Dome C Ice Cores Deuterium Data," IGBP PAGES, World Data Center for Paleoclimatology, Data Contribution Series # 2004–038, NOAA/NGDC Paleoclimatology Program, Boulder CO; and Zhu et al., 2018. The shaded area is shown on figure 9.2.

July at 65° north. A sharp rise in insolation, starting 22,000 years ago, brought the Earth quickly—although somewhat jerkily—out of a glacial cycle. The temperature rose 9°C between 17 and 10.5 ka, or about 0.14°C/century (compared with over 1°C/century at present). The insolation curve started down again at about 10 ka, bringing global temperatures down with it until around 8 ka, at which time there was a surprising (in retrospect, of course) reversal of the temperature trend, with a slow rise until 6 ka, and then generally steady temperatures from then until the industrial period, in spite of the continued drop in 65° north insolation.

Similar data for atmospheric carbon dioxide and methane levels are shown on figures 9.2b and 9.2c. The carbon dioxide level follows the insolation curve upward until about 10 ka and then drops slowly, only to start rising again at around 8 ka. Methane also follows the insolation upward (notwithstanding a large drop during the Younger

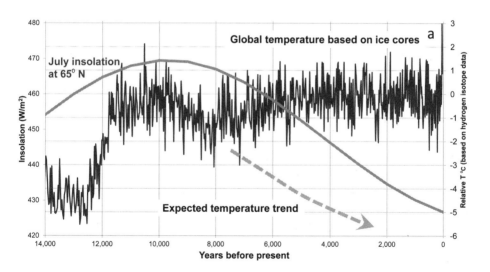

Figure 9.2a. Global temperature over the past 14,000 years, based on Antarctic ice cores, compared with insolation at 65 north4. Based on data in Jouzel, J., "EPICA Dome C Ice Cores Deuterium Data," IGBP PAGES, World Data Center for Paleoclimatology, Data Contribution Series, 2004. Insolation curve based on data from Berger, A., and Loutre, M-F., "Insolation Values for the Climate of the Last 10 Million Years." *Quaternary Science Reviews*, V. 10, pp. 297–317 (Supplement: Parameters of the Earth's orbit for the last 5 Million years in 1 kyr resolution), 1991.

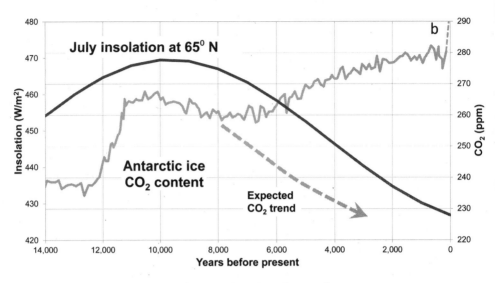

Figure 9.2b. Global atmospheric carbon dioxide over the past 14,000 years, based on Antarctic ice cores, compared with insolation at 65 north. Based on data in Luthi, D., et al., "High Resolution Carbon Dioxide Concentration Record 650,000–800,000 Years Before Present," Nature, V. 453, pp. 379–82, 2008. Source data for insolation curve same as for figure 9.2a.

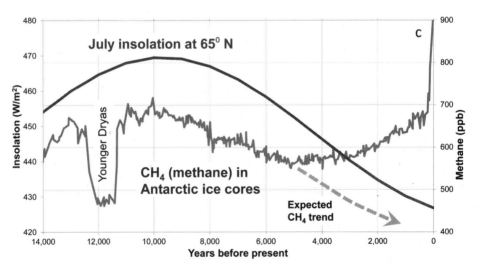

Figure 9.2c. Global atmospheric methane over the past 14,000 years, based on Antarctic ice cores, compared with insolation at 65 north. Based on data in Loulergue, L., et al., "Orbital and Millennial-scale Features of Atmospheric CH_4 over the Past 800,000 Years," Nature, V. 453, pp. 383–6, 2008. Source data for insolation curve same as for figure 9.2a.

Dryas cool period), peaks at around 10 ka, then gradually falls until around 5 ka but then starts to rise after that. It's important to remember in this context that, under natural conditions, methane and CO_2 levels in the atmosphere typically follow and reinforce temperature changes, rather than cause them. That's because, during cooling, atmospheric methane and CO_2 get stored in permafrost while CO_2 gets dissolved into the ocean (as it is more soluble if the water is cool).

The figures above show temperature and atmosphere composition over the past 14,000 years, based on Antarctic ice cores, compared with July insolation at 65° north.[4]

There is no known natural explanation for the failure of the temperature, carbon dioxide, and methane trends to follow the insolation trend down over the past 8,000 years. There is no evidence for a change in ocean currents or massive volcanism or an extraterrestrial impact or for a change in the output of the sun. The only significant change on the Earth during this period was in the number of humans and in their lifestyles.

Agriculture

The earliest evidence of agricultural practices by our ancestors dates to around 17 ka in the southern Levant region (now Israel, Palestine, and Jordan).[5] Farming technology appears to have spread from there to the eastern Fertile Crescent (parts of Iraq and Iran) by 14 ka, to the northern Levant (Syria, Lebanon, southern Turkey) by 12 ka, to the rest of the Fertile Crescent (including Cyprus, Turkey, and Egypt) by 10 ka, and then slowly to the rest of Europe, reaching as far as southern Scandinavia and the British Isles by around 4 ka. Early crops in the Fertile Crescent and Europe included grains (wheat and barley), pulses (lentils, chickpeas, and fava beans), and figs. Domestication of animals started at around the same time, with pigs, goats, sheep, and then cattle. Pasture had to be dedicated to the feeding of cattle and later horses.

Other agricultural crops were developed independently in several other locations. In eastern China, rice farming started around 9 ka,[6] with evidence of a shift to wetland rice cultivation at about 7 ka. Maize and squash were being farmed in Mexico as early as 9 ka.[7] In Bolivia and Peru, wild potatoes had been developed into food crops possibly as early as 8 ka. Other early centers of agriculture include India, the Horn of Africa, and New Guinea.

We don't know a great deal about early farming methods, but it is almost certain that "slash and burn" was an important part of the process and that farming in general was significantly land-intensive. According to University of Maryland landscape ecologist Erle Ellis,

> From my point of view, from my experience—and really pretty mainstream experience—is that obviously they used a lot more land per person in the past. Land was basically free back in the origins of agriculture. There was no land shortage. People used the least labor methods which is burning out the landscape and throwing out a few seeds.... They used large amounts of lands per person to farm.[8]

Soil used in this way would not have lasted long, unless manuring was employed, so after several years, it would have become necessary to burn more natural vegetation and clear more land.

In work published in 2001 and 2003,[9] climate scientist William Ruddiman proposed that human greenhouse gas emissions (primarily from agricultural practices) are responsible for the anomalous climate response of the past 8,000 years. In 2007 he wrote,

> (1) Anthropogenic effects on greenhouse gases and global climate began thousands of years ago and slowly increased in amplitude until the start of the rapid increases of the industrial era. (2) Global climate would have cooled substantially during recent millennia, but anthropogenic greenhouse gas increases countered much of the natural cooling. (3) Had it not been for human interference in the operation of the climate system, ice caps and small sheets would have begun forming in north polar regions. (4) Shorter-term climatic oscillations during the last 2000 years resulted in part from pandemics that caused massive mortality, reforestation, and sequestration of carbon.[10]

Ruddiman's hypothesis is controversial because the global population over the period in question is so much smaller than today's population. There were only about 12 million people on Earth 8,000 years ago. By 5,000 years ago, it had reached 45 million, and then 190 million by 2,000 years ago.[11] Detractors question whether it is possible for a population that small to affect the climate. Ruddiman argues that there are several contributing factors that need to be considered. Firstly, as already noted, early agricultural methods were highly inefficient compared with those used today. Large areas were required to grow relatively little food, requiring extensive destruction of forests to grow any crops. Later on, vast methane-emitting wetlands were created to grow rice. In other words, each person who relied on agriculture for food had a much larger agriculture-related footprint than people do today. Secondly, poor farming practices resulted in the degradation of natural soils and contributed to soil erosion and slope failure, releasing large amounts of carbon that had been stored in the soil. Finally, several climate feedbacks have contributed to the slow warming of the past 8,000 years, including enhanced release of stored carbon dioxide and methane from warmed wetlands and enhanced release of carbon dioxide from warmed oceans.

If we accept Ruddiman's theory, we are forced to ask another question. If a few hundred million people can contribute to a temperature increase of close to 1°C over 8,000 years, simply by living off systematic farming, what is likely to be the ultimate effect of 8 billion people (or the expected peak 10 billion) who are eating food produced by intense farming and are using fossil fuels at a truly alarming rate? Ruddiman has shown that a major component of the preindustrial anthropogenic warming can be attributed to feedbacks, and there is little doubt that the same will apply to postindustrial warming. The longer we wait to control our emissions, the more serious those feedbacks will become.

Fossil Fuels

Starting thousands of years ago, our ancestors used fossil fuels on a very small scale, mostly as coal to produce heat. The significant use of fossil fuels is linked to the industrial revolution in the early 1800s.

As described by Simon Pirani,[12] coal played a pivotal role in two aspects of the machine age: first, when converted to coke it was used for making the iron to produce machines, and second, it was used as the fuel to run the steam engines that powered those machines. Global consumption of fossil fuels from 1800 to the present is depicted on figure 19.3. As shown in the inset for the period 1800 to 1920, coal consumption increased dramatically from 1860 to about 1910 as improvements were made to steam engines and the technology was applied to manufacturing, then to ships and trains, and eventually even to automobiles.

Internal combustion engines were developed in the latter part of the nineteenth century but became widely used only after WWI. This is reflected in the growth of oil between 1920 and 1950. After 1950, oil consumption ballooned as Americans first, and then later everyone else who could afford it, took to the roads with unbridled enthusiasm. This trend was accentuated in North America by the massive develop-

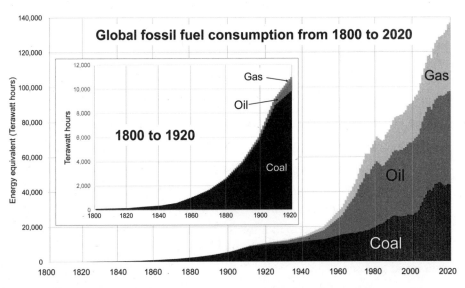

Figure 9.3. Global consumption of fossil fuels over the past 220 years. The inset shows the first 120 years at an expanded vertical scale. Based on data from "Fossil fuel consumption," *Our World in Data.*

ment of the suburbs. These far-flung neighborhoods were not served by the struggling inner-city transit systems, forcing almost every family to have a car, and many to have more than one.

Electrical grids were first established in North America and Europe in the 1910s to 1930s, and the demand for electricity in every corner of our lives soon took off. Much of that demand was first satisfied by coal-fired generation, but gas has been a major contributor in recent decades. In 1971, coal accounted for 40% of electricity production, oil 21%, and gas 13%. In 2018 coal was down a little to 38%, oil way down to only 3%, and gas was up to 23%.[13] Most of the observed growth in natural gas consumption since about 1970 has been from electricity generation.

People

As described above, we can probably attribute the slow warming of our climate over the past 8,000 years to the development and evolution of agriculture, and we can definitely attribute the very rapid warming of the past 150 years to our profligate use of fossil fuels. Both food energy and fossil-fuel energy are also prime contributors to the explosion of the human population. Around 8 ka (~6000 BC), the population was about 12 million (figure 9.5, inset a). This climbed rapidly as agriculture became more universally applied and as methods improved and outputs increased. By 3000 BC, we numbered about 50 million, and by AD 0, we numbered nearly 200 million.

The current period of exponential growth in the population started around 1920 (figure 9.5, inset b), at which time there were just under 2 billion of us. It's probably no coincidence that 1920 is less than a decade after the invention of the Haber-Bosch process for converting atmospheric nitrogen to ammonia fertilizer (a process that now uses about 4% of the world's natural gas supply). A significant part of the post-1920 population growth can be ascribed to the extra food grown as a product of the enhanced fertility provided by synthetic ammonia fertilizer. Of course, many other agricultural advances have also contributed to this growth.

Clean Natural Gas?

Proponents of fossil fuels often tout natural gas as the clean and green alternative to coal and oil. This is true, to some extent, because a molecule of methane (the main component of natural gas) is simpler than a molecule of fuel oil and far simpler than one of coal. Methane is CH_4, or four hydrogen atoms surrounding a carbon atom. Oil is a mixture of more complicated molecules, one of which is octane: C_8H_{18}, or 18 hydrogen atoms surrounding 8 carbons (figure 9.4).

During the combustion of methane, a unit of energy is released for each C-H bond that is broken, and the single freed carbon atom immediately combines with oxygen to make CO_2. So that's 4 units of energy for 1 CO_2 molecule (a 4:1 ratio).

During the combustion of octane, we get 18 units of energy (from 18 C-H bonds), and we make 8 CO_2 molecules. That's 18 units of energy for 8 CO_2 molecules (a 2.25:1 ratio). Coal is even more complex than oil, so the ratio is even lower. According to the IPCC, on average, combustion of gas releases 469 g of CO_2 per kWh of electricity produced, oil 840 g, and coal 1001 g.[14] So natural gas has a lower greenhouse gas footprint than oil and coal, but it still emits CO_2, and has significant climate-change implications.

But that's not all. While refined natural gas is relatively clean from the perspective of impurities, such as sulphur and nitrogen. Oil, even refined oil, is not, and coal is much worse still, so these fuels release a lot more pollution when burned than does gas. Natural gas is cleaner and greener when compared with oil and coal, but that's not saying a lot!

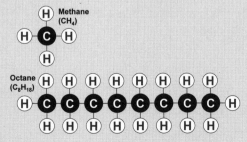

Figure 9.4. The molecular nature of natural gas (as exemplified by methane) and of oil (as exemplified by octane).

From a climate change perspective, there are many problems associated with our vast population, including (but not limited to), the following:

- deforestation for housing, commercial, and industrial construction
- more logging and mining for construction and manufacturing
- more GHG emissions from travel
- more roads and airports to accommodate that travel
- more food production, with GHG emissions at every step of the process
- more consumption of products, leading to more emissions
- more waste to more landfills
- and, of course, more feedbacks

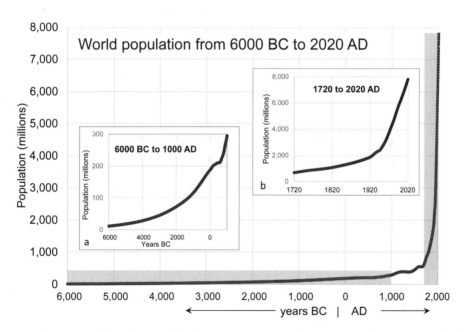

Figure 9.5. Global population between 6000 BC and AD 2020. The insets a and b represent the two shaded areas. Based on data from McEvedy, C., and Jones, R., *Atlas of World Population History, Facts on File*, New York, pp. 342–51, 1978, relying on archeological and anthropological evidence, as well as historical documents such as Roman and Chinese censuses, "World Population Growth," *Our World in Data*.

But of course, it's not just a problem that there are too many people. There is also the problem of what many of those people are doing, such as living in larger houses than ever; driving faster, farther, and more often than ever; taking more frequent and longer trips by air; eating more emission-intense foods; and creating more waste. We'll take a closer look at these lifestyles, and what we can do about them, in chapter 11.

10

TIPPING POINTS

I present multiple lines of evidence indicating that the Earth's climate is nearing, but has not passed, a tipping point, beyond which it will be impossible to avoid climate change with far ranging undesirable consequences.

—James Hansen, 2005[1]

While the high-level climate talks pursue their stately progress towards some ill-defined destination, down in the trenches there is an undercurrent of suppressed panic in the conversations. The tipping points seem to be racing towards us a lot faster than people thought.

—Gwynne Dyer, 2008[2]

IN A LECTURE GIVEN IN 2017, future-thinker Tony Seba showed a photograph of a New York street taken in 1900.[3] The street is full of horses and horse-drawn carriages, with just a single motor car visible. He then showed a photo taken at the same location only 13 years later. The street is full of motor cars, with just a single horse visible. Sometime between 1900 and 1913, New York City crossed a tipping point in its transportation system. To put it into a climate-change context, advertising was the likely forcing mechanism, while convenience, speed, ostentation, social status, and envy provided strong positive feedbacks. The result was cleaner streets of course, and, 107 years later, a serious climate crisis.

Approaching a climate tipping point is a bit like walking toward the edge of a cliff in the dense fog. You may know it's out there, but you don't know how close you are, and by the time you've crossed it, it is probably too late. To be more specific, a climate tipping point occurs when a region (or the Earth as a whole) crosses an irreversible threshold in its ecological or physical state, one that can lead to larger changes. The term "irreversible" is important here because it means that it isn't possible for us, or any natural process, to undo the change, and that it probably won't change back naturally, at least not on a human time scale. Going back to 1913, by then there was nothing that the horse-and-carriage industry could do to convince New Yorkers to abandon the motor car.

As I write this in the fall of 2020, wildfires are raging along the western edge of North America. The first, third, fourth, fifth, and sixth largest fires in California history are burning right now, consuming a record area of forest, over 17,000 km², more than doubling the old record of 8,000 km² set only two years ago.[4] To date, in 2020, this area is equivalent to the combined area of Los Angeles, San Diego, Houston, Indianapolis, Dallas, Nashville, Memphis, Jacksonville, Oklahoma City, Phoenix, San Antonio, Fort Worth, and Louisville. In Oregon, five entire towns, with a combined population of over 11,500, have been razed to the ground within the last few weeks. A total of 7,500 structures have been destroyed in California, Oregon, and Washington, and 37 people have died.[5] Here in British Columbia, there are few fires burning, but we are shrouded in an apocalyptic smoke, reportedly from fires in Oregon and Washington. Figure 10.1 shows that although 2020 isn't an exceptional fire year in the US as a whole (at least not yet), there has been a roughly threefold increase in the area consumed by fires since the 1980. Half-way around the world, an area of 318,000 km² (about the size of Poland) has been burned in Siberia, and earlier this year 186,000 km² was burned in Australia; both are historical records.

According to the U.S. Global Change Research Program: "Increased warming, drought, and insect outbreaks, all caused by or linked to climate change, have increased wildfires and impacts to people and ecosystems in the southwestern US. Fire models project more wildfire and increased risks to communities across extensive areas."[6] Another recent study has shown that autumn precipitation in California has decreased by 30% over the past 40 years and the average temperature has gone up by 1°C, resulting in a doubling of the number of days that are ideal for wildfires to start.[7] California and neighboring states are not getting less precipitation overall, but the differences between wet winter and dry summer/autumn periods are becoming more extreme as a direct result of climate change.[8]

We need to consider whether this multiple-record-breaking wildfire year is the harbinger of an approaching tipping point or possibly one that has already been crossed. Wildfires have a number of positive feedback effects that can lead to larger climate changes, and there are also ways in which wildfire damage can be irreversible. The key positive feedback from fires is the loss of the carbon capturing capacity of the vegetation that gets burned. This results in higher carbon dioxide levels overall and therefore more warming (although

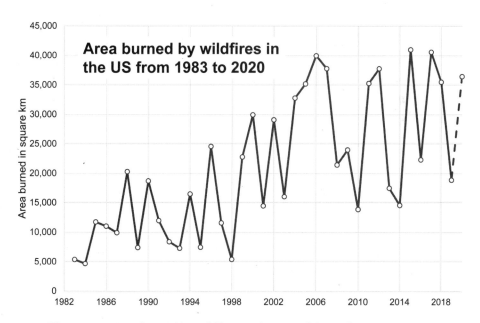

Figure 10.1. Area burned by wildfires in the United States from 1983 to 2020. (The dashed line indicates that the final number for 2020 is not available at the time of writing.) Based on data from the National Interagency Fire Centre, nifc.gov/fireInfo/fireInfo_statistics.html, October 2020.

this may be mitigated if the lost vegetation starts to recover quickly). Another positive feedback is from the destabilization of soil. Areas affected by wildfires are highly vulnerable to soil erosion and slope failure, and both of those add to the emissions of carbon dioxide and methane. They also make it harder for forests to regrow. These effects exacerbate global climate change and so enhance the likelihood of future wildfires in many areas, and that alone could push some of them over a tipping point.

But there is also a possibility that forests will not regrow at all in some areas that have suffered fires recently, and this is why the changes may be irreversible. The mature ecosystems that have been destroyed in recent wildfires started growing many decades or even centuries ago under cooler and wetter conditions. They have been able to survive in the current climate because they are well established, but there is a good chance that similar plant communities will not be able to get a foothold in this climate. This is exacerbated by the fact that wildfires can also lead to local changes that make regeneration more difficult. For example, while the loss of forest cover increases albedo and that has a cooling effect, it also reduces evapotranspiration, which has a warming effect.[9]

Indeed, there is widespread evidence that ecosystems in many of the relatively dry parts of the western United States are struggling to recover from wildfires.[10] The authors of one study state,

> At dry sites across our study region, seasonal to annual climate conditions over the past 20 years have crossed these thresholds, such that conditions have become increasingly unsuitable for regeneration. High fire severity and low seed availability further reduced the probability of postfire regeneration. Together, our results demonstrate that climate change combined with high severity fire is leading to increasingly fewer opportunities for seedlings to establish after wildfires and may lead to ecosystem transitions in low-elevation ponderosa pine and Douglas-fir forests across the western United States.[11]

It is possible that this represents a series of small tipping points happening in many different places, and at different times, all across the western US (and likely western Canada as well). It is also possible that these may merge into a major tipping point in the larger ecosystem.

Past Tipping Episodes

Before we start examining other potential present and future tipping points, it's worth taking a quick look back at some past episodes when the climate changed abruptly from one state to another. One of these, described in chapter 6, is the Atlantic salinity oscillator (see figure 6.5 and relevant text). This phenomenon is related to changes in the surface and deep-water circulation system in the Atlantic Ocean, also known as thermohaline circulation (THC). The THC brings warm salty Gulf Stream water from the tropics toward Greenland, Iceland, and continental Europe. When that salty water reaches the far north Atlantic, it cools and becomes dense enough to sink, and then it flows back south at depth. The warmth of the Gulf Stream keeps western and northern Europe relatively mild and also contributes to the melting of ice on Greenland and Iceland (figure 10.2a). Under this scenario, lots of fresh water from melting glaciers flows into the north Atlantic, diluting the Gulf Stream. If this dilutes the water enough, it won't sink, even when it gets cold, and the THC will slow down. That will likely lead to a short circuit in the THC, as shown on figure 10.2b, resulting in cooling in the north, less glacial melting, and then eventual reestablishment of a strong THC.

During the last glaciation, this process happened repeatedly on a time scale of about 1,500 years. In one of these cycles, starting 35,500 years ago, THC strengthening caused the Greenland temperature to rise by almost 7°C in just 350 years, and then THC weakening led to a similar drop in temperature over the next 1,000 years. This represents a tipping point because it was a dramatic climate change (at least for Greenland, Iceland, and continental Europe) and because it was irreversible, at least on a human time scale.

Another significant tipping point occurred about 55.8 million years ago at the boundary between the Paleocene and the Eocene

Epochs. Over a period of a few thousand years, sea surface temperatures rose by between 5° and 8°C,[12] and stayed that high for about 150,000 years.[13] The Paleocene-Eocene Thermal Maximum (PETM) led to a major extinction of marine microorganisms. It is not known what caused the initial temperature increase at the end of the Paleo-

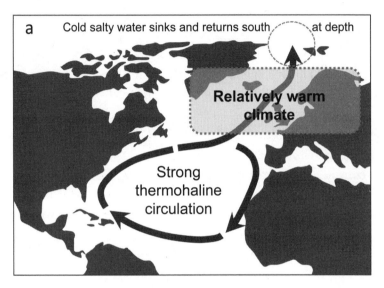

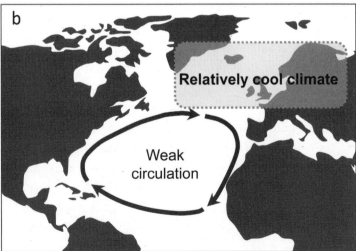

Figure 10.2. Salinity oscillation in the north Atlantic basin during the last glaciation

cene, but the tipping point in this case came when sea floor methane hydrate deposits started breaking down and emitting large volumes of methane into the ocean and the atmosphere.

Methane hydrate is a combination of ice and methane in which the ice molecules form a cage around a methane molecule (figure 10.3). It is a solid that looks like regular ice, but it will burn if you put a match to it, and it is a very compact way of storing methane. When 1 cubic centimeter of methane hydrate breaks down, 164 cubic centimeters of methane gas will be released. It is currently present (and has also been in the past) in vast deposits within sediments at and just below the seafloor. Seafloor methane hydrate is kept stable by low water temperatures and high pressure, but it can become unstable if the deep ocean water warms by a few degrees. At present there is more energy stored in sea floor methane hydrate deposits than all of the conventional coal, oil, and gas ever produced plus all of that still in the ground.

Methane hydrate is also present at depths of tens to hundreds of meters within sea floor sediments in polar regions and within

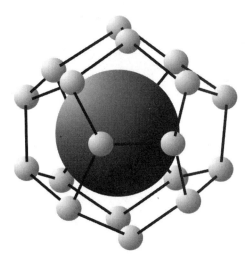

Figure 10.3. A typical methane hydrate structure. The dark gray sphere represents a methane molecule (CH_4), and the lighter spheres represent water molecules (H_2O).

permafrost deposits both on land and offshore in the Arctic Ocean. Those offshore hydrates formed on land during the Pleistocene glaciations and have since been flooded by post-glacial sea-level rise.[14]

It takes a very long time for deep ocean water to warm up during climate change, and so climate scientists are not predicting that the deeper sea floor methane hydrate deposits will start to break down within the next century. On the other hand, methane hydrates in permafrost on the floor of the shallow parts of the Arctic Ocean, or on land in polar regions, will likely be vulnerable sooner than the deep deposits. In fact, they are vulnerable now.

Potential Future Tipping Points

There are numerous aspects of the Earth's climate system that could develop into tipping points in next few centuries, decades, or even the next few years. Some of the better-understood examples are shown on figure 10.4. These are described in the following pages.

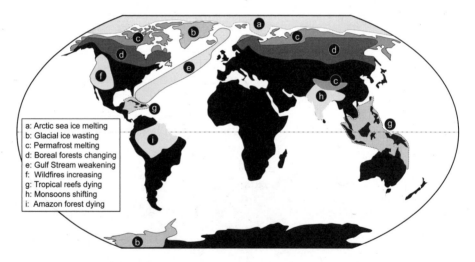

a: Arctic sea ice melting
b: Glacial ice wasting
c: Permafrost melting
d: Boreal forests changing
e: Gulf Stream weakening
f: Wildfires increasing
g: Tropical reefs dying
h: Monsoons shifting
i: Amazon forest dying

Figure 10.4. Locations of some of the regions where climate tipping points are imminent or may become so within decades. Based partly on information in a February 2020 article at *Carbon Brief* by Robert McSweeny, accessed September 2020; on information in Lenton et al., "Tipping Elements in the Earth's Climate System," *Proceedings of the National Academy of Sciences*, V. 105, pp. 1786–93; and Potsdam Institute for Climate Research, pik-potsdam.de/en /output/infodesk/tipping-elements.

Arctic Sea-ice Melting

The floating sea ice in the Arctic Ocean typically reaches its lowest areal extent in September of each year. The area in September 2020 was second only to the all-time minimum (reached in 2012), and the typical minimum area in the 2020s is about one-half of what it was in the late 1980s and early 1990s (figure 10.5). Equally significant, is that the average thickness of the ice is also down to about one-half of what it was in the early 1990s,[15] and so the volume of ice is now around one-quarter of the volume just 30 years ago. The thinner ice that now remains is increasingly vulnerable to melting.[16] In other words, we are close to a tipping point in the Arctic sea-ice system, from the state where there has been year-round sea ice for tens of thousands of years to one where there isn't.

The observed changes already have significant implications for the Arctic climate and ecosystems but increasingly also for the rest of the Earth. According to Cambridge climate scientist Peter Wadhams, there are some major implications of what he calls the Arctic "death spiral,"[17] as follows:

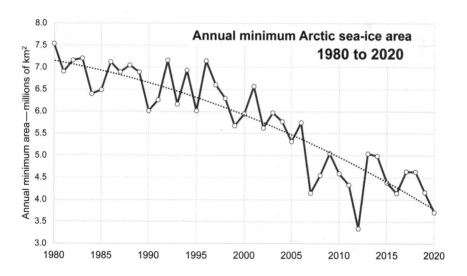

Figure 10.5. Arctic Ocean September sea-ice areal extent from 1979 to 2020. (The dotted line is a second-order polynomial curve fit to the data.) Based on data from the National Snow and Ice Data Centre, University of Colorado at Boulder, nsidc.org/data/seaice_index/archives.

- The most obvious implication of melting sea ice is the albedo effect because open water absorbs much more solar energy than ice or snow. This is leading to both regional (Arctic) and global temperature increases.
- There are large stores of methane and carbon dioxide in permafrost, both on land in the Arctic and on the seafloor in areas that were flooded at the end of the last glaciation. As the Arctic temperature rises, these GHGs are being released to the atmosphere, leading to enhanced warming everywhere.
- Glaciers on Greenland and many Arctic islands are melting faster because of the local heating, also contributing to albedo-related warming, to release of GHGs, and, of course, to sea-level rise.
- Because of higher temperatures and a greater area of open water, the Arctic air has more water vapor than normal. Since water is a GHG, that is contributing to warming.
- Warming of Arctic lands leads to warming of Arctic rivers, which are now bringing even more warmth to the Arctic Ocean.

The decline of Arctic sea ice has significant implications for some of the other potential tipping points that are discussed below, especially for the melting of ice on Greenland and for the stability of permafrost on land in Arctic regions and on the shallow seafloor of the Arctic Ocean offshore from Siberia.

Lenton and coauthors have predicted that the tipping point for loss of Arctic sea ice is between 0.5° and 2°C of global warming.[18] Since we have already seen more than 1°C of post-industrial global warming, we have either passed that threshold or could within another decade or two.

Loss of Glacial Ice in Greenland and Western Antarctica

A warming climate has led to a significant decrease in the volume of glacial ice in Greenland and Antarctica, and there are indications that the rates of ice loss in these areas could accelerate rapidly in the future. As shown on figure 10.6, the overall mass of Greenland ice did not change significantly from 1972 to 1986, but there has been a dramatic loss of ice in the 32 years from 1986 to 2018, and especially since 2000, amounting to a total loss of about 1,500 Gt, or more than three times the volume of Lake Erie.

During the same time period, the loss of ice from Antarctica has been approximately 5,000 Gt,[19] with fully half of that coming from the West Antarctic Ice Sheet, which makes up less than 20% of the area of Antarctica.

Ice sheets in both Greenland and West Antarctica are losing mass through melting at surface, which accounts for about 34% of the mass loss in Greenland and 10% of the mass loss in West Antarctica, and through glacial flow into the ocean, which accounts for the majority of the mass loss, especially in West Antarctica. There are two main causes for enhanced flow of Greenland and Antarctic glaciers: one is the transfer of meltwater from the ice surface to the ice-bedrock

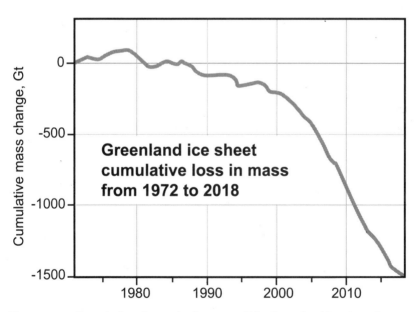

Figure 10.6. Cumulative change in the mass of the Greenland ice sheet from 1972 to 2018 (a giga-tonne, GT, is 1012 kg). Based on data in Mouginot, J., et al., "Forty-six Years of Greenland Ice Sheet Mass Balance from 1972 to 2018," *Proceedings of the National Academy of Sciences*, V. 116, pp. 9239–44, 2019.

interface, where it acts as a lubricant for ice flow; and the second is the melting of the submerged leading edges of glaciers by relatively warm sea water, which promotes ice advance. Melting and collapse of the floating ice shelves adjacent to the glacier fronts also promotes ice advance.

The main implication of rapid melting on Greenland and Antarctica is sea-level rise, and although the rates are still small at present, millimeters per year, it is almost certain that they will increase significantly in coming decades. Melting on Greenland also has implications for dilution of the Gulf Stream and the consequent slowing of the Atlantic thermohaline circulation, as described above.

Melting and Destabilization of Permafrost

The Batagaika Crater is a giant gaping hole in the Yakutia region of Siberia. It is 60 m deep with an area equivalent to 111 soccer fields (780,000 m²), and it is growing by the equivalent of about 3 soccer fields a year. It started forming when the trees in the surrounding region were cut in the 1960s. That allowed the permafrost to start thawing, and now there is nothing that can stop it.[20] The permafrost that is collapsing is rich in carbon, and that carbon is being released into the atmosphere in the form of carbon dioxide and methane.

Permanently frozen soil exists in non-glaciated areas at high latitudes or at high elevations where the mean annual temperature is consistently below 0°C. It is called permafrost if it persists for at least two years, although most permafrost has existed at least since the last deglaciation (about 12,000 years). Permafrost conditions exist on about 25% of the land in the northern hemisphere. The greatest areas are in Arctic regions of Russia, Canada, and Alaska, but there is also extensive permafrost on the Tibetan Plateau and adjacent Himalayan Mountains, and less extensive areas on other northern hemisphere mountain ranges.

Batagaika Crater is perhaps the most striking example of permafrost collapse and disintegration, but there are thousands of similar sites around the Arctic vast areas where permafrost is simply melting without collapsing. It has been estimated that the frozen materials that make up permafrost hold twice as much carbon as is currently in the atmosphere,[21] and while it will take centuries for this to be released, the rate of breakdown is accelerating, and the feedbacks, which include higher GHG levels and decrease in albedo of degrading sites, are strongly positive. Furthermore, the Arctic region is warming faster than most other parts of the world, in part due to the loss of Arctic sea ice.[22]

A significant part of Siberia was not glaciated during the Pleistocene, and that included a large area that was land at the time but is now flooded by the Arctic Ocean because sea level has risen by over 100 m since deglaciation. Much of the East Siberian Sea has water depths of less than 50 m and is underlain by permafrost that formed on land. There is evidence of significant release of methane from these areas,[23] likely because they now experience warmer temperatures on the seafloor than they did on land.

Boreal Forest Changing

The boreal forests (or taiga in Europe and Asia) comprise 29% of the world's forests by area and represent a disproportionately large portion of biologically sequestered carbon. As shown on figure 10.4, boreal forests extend across all of northern Russia and into Scandinavia and across all of northern Canada and into Alaska. They are dominated by conifers (especially spruce, pine, and larch) with some deciduous trees (especially birch). There is significant overlap between the boreal forest and the permafrost regions, with extensive permafrost in the more northerly parts of the boreal forests.

Because northern regions are warming faster than the rest of the Earth, the boreal forests are becoming increasingly affected by rising temperatures. The southern margins of the biome[24] are facing dieback because of warmer temperatures and drought, and wildfire damage is an increasing issue,[25] while in the northern parts, shrubsized vegetation is advancing out into the tundra. This greening along the northern edge is not a positive change from a climate perspective because it makes those areas darker than before, so that they absorb more heat and that leads to warming and greater degradation of permafrost.

According to Lenton and coauthors, the global temperature increase threshold for dieback of the boreal forest is 3°C, although there is significant uncertainty about that number,[26] and it is likely that increasing wildfire activity could play an accelerating role.

Gulf Stream Weakening

As described above under "Past Tipping Episodes," the strength of the thermohaline circulation (THC) in the Atlantic Ocean, which brings warmth to northwestern Europe, is dependent on cold salty water sinking in the far north and flowing south at depth. This system has tipped numerous times in the past, the likely mechanism, at least in part, being dilution due to strong glacial melting on Greenland and northern Canada.

There is multiple evidence that the THC could be heading toward a transition soon. For example, there is cooling of north Atlantic water (see figure 6.6 and the related text); there is increased input of freshwater from acceleration of melting on Greenland; and there are measured reductions in the salinity of the north Atlantic.[27]

Lenton and coauthors suggest that the THC will cross a tipping point this century.[28] When this happens, there will be cooling in western Europe, although that will likely be offset by warming elsewhere.

Wildfires Increasing

Figure 10.1 illustrates that there has been a clear increase in the area consumed by wildfires in the US over the past four decades. It has increased from an average of about 12,000 km² in the 1980s and '90s, to about 25,000 km² in the first decade of this century, to 30,000 km² in the past decade.[29] As already noted, Canada has also seen an increase in the area consumed by wildfires in the past several decades.[30]

As described at the start of this chapter, there is evidence that tipping points have already been crossed in some areas of western North America, where plant communities that developed decades ago are able to persist in the current climate only because they are mature, and when they are destroyed by fire, they do not necessarily grow back. These localized tipping points are feeding back into warming the climate, contributing to fires and to tipping points in other areas, such that, within several decades, large parts of North America will look very different.

And it's not just North America that is changing in this way because of runaway fire activity. Similar things are happening in Australia, Russia, Indonesia, and the Amazon.[31]

Tropical Reefs Dying

Damage to corals within tropical reefs occurs when the water temperature exceeds their range of tolerance and the symbiotic relationship between the coral structures and the algae (zooxanthellae) living within their tissues breaks down. The result is bleached—and therefore likely dead—coral, and the ecosystem implications are huge.

As shown on figure 10.7, there has been a significant increase in the incidence of coral reef bleaching in recent decades. During the first part of the period shown (1980 to 1995), significant severe bleaching (affecting >30% of the corals at a site) was restricted to El Niño years and never exceeded 25% of the sites studied. During the remainder of the period (1996 to 2016), significant bleaching has also taken place in non-El Niño years, with more than 30% of sites affected in some years and almost 50% affected in the most recent year of the study (2016). Terry Hughes, the lead author of the report on which figure 10.7 is based, has recently noted that there was also widespread severe bleaching in 2017 and 2020, the latter being second only to 2016 in extent.[32] Neither 2017 nor 2020 were El Niño years.

There is significant concern that coral reef communities are close to a tipping point. A 2018 IPCC report states,

> Warm water (tropical) coral reefs are projected to reach a very high risk of impact at 1.2°C, with most available evidence suggesting that coral-dominated ecosystems will be non-existent at this temperature or higher (*high confidence*). At this point, coral abundance will be near zero at many locations and storms will contribute to "flattening" the three-dimensional

structure of reefs without recovery, as already observed for some coral reefs.[33]

The global temperature increased by 1.0°C between 1950 and 2020 and is on track to exceed 1.2°C by 2030. Coral reefs are vital to tropical marine ecosystems, and their decline will have far-reaching implications. They are also critical sites of carbon sequestration, and so any productivity reductions will have global climate implications.

Monsoon Patterns Shifting

In simple terms, a monsoon is a regional wind pattern that results from summertime heating of land surfaces, while nearby ocean surfaces remain relatively cool. That leads to rising air over the land and brings moist air into land areas from oceans, resulting in enhanced precipitation. The best-known example is the South Asia Monsoon, which affects Pakistan, India, Bangladesh, and other countries in

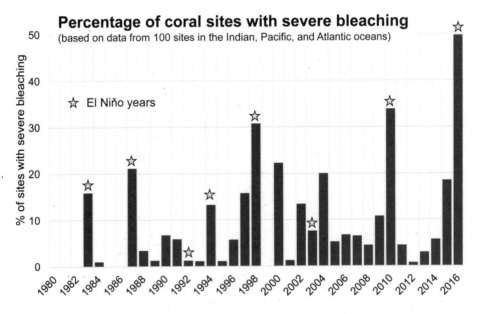

Figure 10.7. Percentage of coral sites with evidence of severe bleaching in the tropical Indian, Pacific, and Atlantic Oceans from 1980 to 2016. Based on data from Hughes, T., et al., "Spatial and Temporal Patterns of Mass Bleaching of Corals in the Anthropocene, *Science*, V. 359, pp. 80–3, 2018.

Southeast Asia between June and September of each year. The monsoon accounts for about 80% of the precipitation in India, and so is critical to the prosperity of that country and of neighboring countries.

While it is expected that a warming climate should strengthen the South Asia Monsoon (because land surfaces warm faster than oceans and because there is more moisture in the air in general), there is evidence that rainfall amounts in India have declined by about 10% since 1950. One possible explanation is that the monsoon is weakening because air pollution is leading to greater cooling over land than over the ocean.[34] Although a tipping point in the South Asia Monsoon is probably many decades away, the issue is an important one because so many people could be affected.

Amazon Forest Dying

Like many forests, the Amazon is the author of its own success or failure. As much as 30% of the precipitation that falls in the region is recycled water from evapotranspiration generated by the vegetation; therefore, any loss of forest cover will result in a loss of precipitation, and that could then spiral into a major change to the ecosystem. Dieback of the Amazon forest has been predicted for a global mean temperature increase of 3° to 4°C because that would lead to changes to the El Niño cycle that would result in strong drying in the region.[35] Although a 3° to 4°C increase is many decades away, even under the worst climate scenarios, and might not be reached for over a century under the most optimistic scenarios,[36] there are two key factors that could accelerate the tipping point. One is intentional deforestation, which has been happening at an increased pace in recent years, and the other is wildfire, which is partly a result of regional warming and drying but is also related to deforestation because much of that is achieved through burning.

Summary

The nine tipping points described above are listed in table 10.1 along with a summary of their likely regional and global implications and an indication of the possible timescale of tipping. It is evident that that time may be very short (or even passed) for Arctic sea ice loss, biome shift in western North America related to wildfires, and the death of tropical reefs, while other tipping points are decades or possibly centuries away.

While some tipping points may have only regional implications, others could affect the climate on a global scale and could lead us into runaway climate change. It's likely that permafrost melt could do this alone, whereas some other processes could do it if they were part of a cascading chain where one type of dramatic change contributes to others.

So, let's pretend that it's you stuck on a plateau in fog so thick you can barely see your hand in front of your face. You know there's a cliff edge nearby, but you don't know how far away. If you stumble over the cliff, you might die, or you just might be stuck in a really bad (and irreversible) situation, injured, with nobody to rescue you. You have to make a choice. Continue to wander aimlessly, hoping for the best,

Table 10.1. Summary of some important tipping points, their regional and global implications, and their possible timing

Tipping point	Regional implications	Global implications	Timescale
Arctic sea-ice loss	Warming, glacier & permafrost melt	GHG increase, warming	now (?)
Glacial ice-sheet melt	Minimal	Sea-level rise	many decades
Permafrost melt	Biome change	GHG increase, warming	decades
Boreal forest shift	Biome change	GHG increase, warming	many decades
Gulf Stream change	Europe cooling	Warming (?)	decades
Wildfire increase	Biome change, less precipitation	GHG increase, warming	now (?)
Tropical reef death	Ocean biome change	GHG increase, warming	now (?)
Asia monsoon change	Biome/agriculture change	Limited	decades
Amazon forest death	Biome change	GHG increase, warming	decades

or stop and wait for the fog to clear. The choice is pretty clear, I think, but you have to make it now.

In a similar way, we have now brought the Earth very close to a climate tipping point. We just don't know how close because this is new terrain for us. We can continue to stumble around and hope for the best, or we can make some major changes that should bring us back from the brink. Of course, deciding to significantly change how we live is much, much more complicated than deciding to stop blundering around in the fog near to the edge of a cliff, but the consequences of making the wrong decision are dire in both cases. In one, a single person faces a dangerous fall over a cliff. In the other, an entire civilization (plus much of nature) faces collapse. The choice is clear, and we have to make it now.

11

WHAT NOW?

Our problems are man-made; therefore, they may be solved by man. No problem of human destiny is beyond human beings.

— John F. Kennedy, as part of an address titled "Towards a Strategy of Peace" given at the American University, Washington, DC, June 10, 1963

PRESIDENT KENNEDY was talking about peace back in 1963, not about climate change, but the message is the same. We got ourselves into this mess, and we are capable of getting ourselves out, if we put our minds and our hearts into it. Had President Kennedy lived for another 30 or 40 years, I like to think that he would have been on the side of taking serious action to solve the problem of climate change. We all need to be on that side because if we don't take serious action soon, the consequences will become increasingly devastating and deadly, and extraordinarily expensive.

So, what do we need to do to avoid a stumbling over the edge of the cliff in the fog? To answer that question, we need first to understand which of our current behaviors are the most significant contributors to the problem. There is a persistent emphasis on "we" in this chapter because we as individuals need to own this problem. We cannot place all of the blame on governments, other countries, corporations, farmers, or other people. We cannot expect others to make changes if we are not willing to make some important and difficult changes ourselves, and we can no longer afford to behave like toddlers by say-ing "If they won't change, why should I."

Although "we" all need to take personal responsibility for climate change, our governments also have a critical role to play in solving this problem. That role includes acknowledging that we have a prob-lem in the first place, setting aggressive short- and long-term climate goals, establishing incentives for consumers to help them make climate-friendly choices, and passing regulations that will force corporations to reduce their emissions, to develop climate-friendly products, and to stop trying to coerce consumers into making climate-wrong decisions.

The Sources of the Problem

The sources of the greenhouse gases that are contributing to climate change are summarized on figure 11.1. The main culprit, of course, is

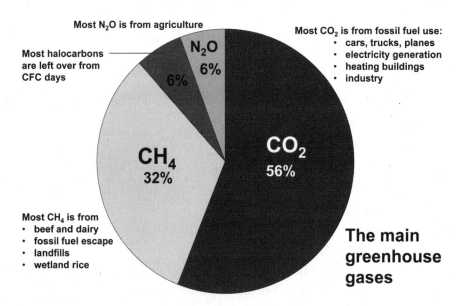

Figure 11.1. Contributions to climate change and their sources. From data in *Climate Change 2013: The Physical Science Basis*, IPCC, 2013. Contribution of Working Group I to the Fifth Assessment Report of the Intergovernmental Panel on Climate Change, Cambridge University Press.

carbon dioxide (at 56%), and the things that we do to produce it are as follows:

- driving cars and flying in planes
- getting stuff manufactured and then brought to us in ships, trucks, and planes
- generating electricity with coal and gas (and then using that electricity for air-conditioning and heating, water heating, appliances of every imaginable type, and lighting)
- heating buildings
- manufacturing things through industrial processes (e.g., using coal to make steel or natural gas to make nitrogen fertilizer).

The proportions of these contributors to our carbon dioxide emissions vary a lot from place to place—depending mostly on how our electricity is generated—and from person to person—depending on how we live our lives.

The second most important GHG is methane, at 32%. As we'll see, most of our anthropogenic methane comes from cows,[1] and so our significant consumption of beef and dairy is a big problem. A lot of methane is unintentionally (or carelessly) released during the production and processing of fossil fuels, some comes from landfills and sewage, and most of the rest is from wetland rice production.

Halocarbons are gases like the chlorofluorocarbons (CFCs).[2] Although their production and use has dropped dramatically since the 1989 Montreal Protocol, they are long-lived molecules, and there is still enough left in the lower atmosphere to contribute to about 6% of warming.

Nitrous oxide (N_2O)[3] accounts for 6% of warming. It is primarily produced by agricultural operations, especially by fertilization (including lawns and golf courses) and as a product of manure. Some comes from fossil fuel combustion.

Since a total of 88% of GHG warming results from carbon dioxide and methane, the activities that result in emission of those gases are what we need to focus on to get climate change under control.

Reducing Carbon Dioxide Emissions

As summarized above, most of the carbon dioxide that is emitted, either by us or on our behalf, is related to transportation, electricity generation and heating, and manufacturing. Here are some effective strategies for controlling those emissions.

Transportation

Transportation is number one, and this is where we, as individuals who make choices, can have the biggest impact. Our first and most important choice is about deciding how to get from one place to another. We could walk, bike, or take a bus, but most of us the just get in the car and drive, without even thinking about it. For people who are determined to drive (or don't have the option not to), the choices come down to what we drive, how often we drive, and even how fast we drive.

So far, we're not doing well when it comes to *what* we drive. In 2019, the best-selling vehicle in the US was a pickup truck (Ford F-150), second was another pickup (Chevrolet Silverado), and third (you guessed it) another pickup (Ram).[4] Small cars accounted for only 10% of sales. The situation is no better in Canada, where 2019 sales of SUVs and pickup trucks outnumbered sales of medium and small cars by three to one.[5]

But there is hope for change, and it's coming quickly. As shown on figure 11.2, sales of plug-in electric vehicles are now quite high in several countries, especially Norway, Iceland, Netherlands, and Sweden (and also in California at 8% of sales in 2019[6]). And yes, there are electric pickups on the horizon too.

In fact, Norway—at 56%—has reached a tipping point for electric vehicles, just as New York City did for motor cars sometime around 1910 (see chapter 10). Norway has achieved this with some popular incentives, including

- no annual road tax on electric vehicles
- 50% saving on ferry fares
- at least 50% saving on parking fees
- access to bus lanes

- corporate vehicle tax reduction reduced to 40%
- no purchase or import taxes
- exemption from 25% value-added tax[7]

These incentives were established so that Norway could meet its commitments under the 2015 Paris Agreement, and that is an example of a government taking a significant initiative for change that has encouraged individual citizens to make their own changes. The incentives are like climate forcings, and the feedbacks are much lower operating and maintenance costs, along with social status.

In Norway, Iceland, and Sweden, there is an additional advantage to the use of electric cars because almost all of the electricity used in these countries is fossil-fuel free. Although that's not the case in the Netherlands or Portugal, where most of the electricity is currently produced using gas and coal, the increasing adoption of electric cars is still a big step in the right direction.

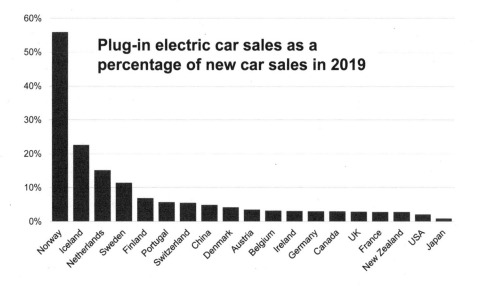

Figure 11.2. The proportion of plug-in electric cars as a percentage of new car sales in several countries. Based on data at "Electric car use by country," *Wikipedia.*

Excuses for continuing to drive a fossil-fuel car are fading fast. The ranges of electric vehicles are getting better, their prices are coming down (especially with generous incentives in many jurisdictions), and the infrastructure for charging is improving all the time. Several countries and states/provinces have now put deadlines on the sale of non-electric vehicles. In Germany, Ireland, Netherlands, and Norway, sales of new fossil-fuel vehicles will be banned after 2025; in the UK, India, and Israel, it is 2030, and in Canada, Spain, and France, it is 2040. Bans in California and Quebec will take effect in 2035.[8]

How often, how far, and how fast we drive are different types of choices. Notwithstanding the implications of COVID-19 (which are discussed below), many people don't have the choice not to drive to work every day because there is no viable public transit, but most of us do have choices about other types of driving. In fact, in the US only 15% of daily trips are made for commuting, while 45% are for shopping and 27% are for social and recreational reasons.[9] Some of the

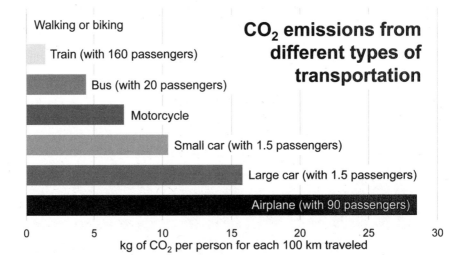

CO$_2$ emissions from different types of transportation

Walking or biking
Train (with 160 passengers)
Bus (with 20 passengers)
Motorcycle
Small car (with 1.5 passengers)
Large car (with 1.5 passengers)
Airplane (with 90 passengers)

0   5   10   15   20   25   30
kg of CO$_2$ per person for each 100 km traveled

Figure 11.3. Emissions from different modes of transportation, in kilograms of CO$_2$ per person per 100 km of travel. Based on data from the European Environment Agency, eea.europa.eu/media/infographics/co2-emissions-from-passenger-transport/view, accessed October 2020.

better choices we can make include walking or cycling for short trips, using transit, planning carefully and combining trips, and rethinking some of our discretionary car trips.

For longer trips, most North Americans have a choice between driving and flying. Trains are a third option in some areas, but much more so in Europe and parts of Asia, where many of the trains are fast enough to compete with air travel. The GHG implications of different types of travel are summarized on figure 11.3. All of the bar lengths on that figure depend on a number of factors, such as the size and type of vehicle, the kind of fuel it uses, the speed it is doing, and, of course, how many people are on board. For air travel, the lengths of the individual hops in a multi-leg journey also matter because taking off requires a lot more fuel than cruising at altitude. The number of passengers shown in the examples on figure 11.3 are based on typical load factors, although they can obviously vary widely. The closer to capacity, the greater the efficiency.

The general message of figure 11.3 is that, from a climate perspective, taking a train or a bus is a much better option than driving, while driving is generally better than flying. Of course, flying may be the only available option for many intercontinental journeys.

Electricity Generation and Heating

We have relatively little control over how our electricity is generated, except to lobby governments and power companies. But some of us have the option of taking things into our own hands by generating our own electricity. The conditions for that have never been better, and they will continue to improve. In most jurisdictions in North America, the levelized cost of energy (LCOE) from residential solar PV installations[10] is now comparable to, or even lower than, the cost of purchasing electricity from the grid. This even applies in most places with less-than-ideal solar resources, like New York or Vancouver. The LCOE for solar will almost certainly continue to drop over the coming years, while the cost of grid electricity will almost certainly continue to rise.

We do have some control over how much electricity we use, and in that regard, the systems to focus on are heating and cooling, hot water, and appliances (especially washing and drying clothing).

From a climate perspective, heating with electricity is generally preferable to heating with gas or oil, but if the electricity you are using is generated using fossils fuels, it doesn't help much. Electric heating with a heat pump is about three times more efficient than heating with baseboards, and so that helps no matter how your electricity is made. It will also save you money.

The most important point is to avoid heating or cooling more than necessary; this could include turning the thermostat down a few degrees in winter and up a few degrees in summer, not heating or cooling rooms that aren't being used, and not heating or cooling at times when the home is empty. Opening or closing blinds to let the sun in—or keep it out—can help, and so can planting shade trees, if that is an option. Trees have the added benefit of sequestering carbon

Washing machines and tumble driers are two of the most important tools for freeing people (especially women) from the drudgery of domestic work, but they also make it easier for us to throw our clothes into the laundry rather than putting them away to be worn again. Perhaps it's too easy, and that can be a problem for the environment and for climate change.

Washing clothes in cold water, running the dishwasher only when it is full, and taking shorter showers are good ways to reduce the climate cost (and other costs) of hot water.

Manufacturing

Making the stuff we like to have requires farming and mining (or extraction of fossil fuels), manufacturing, packaging, and transportation. All of those steps take energy, and most of that energy produces GHG emissions. Craig Jones and Geoffrey Hammond of the University of Bath in the UK have created a database cataloging the important GHG emissions related to a wide range of materials that end up in the things we buy (and build). The Inventory of Carbon and Energy[11] can be used to determine what it costs us—in GHG emissions—to make things, and we can also add in what it costs us to operate them. An example of this is provided on figure 11.4 for some typical household appliances. These are just estimates, of course, because there are hundreds of different types of laptops and refrigerators that are manufactured in different ways in different places, and then used in areas with differing electrical generation types. But the important point is that each type of appliance has a different combination of embodied emissions (emissions from the energy used for resource extraction and manufacturing) and use emissions. Electronic devices tend to have high embodied emissions but relatively small use emissions. White goods are not as emission-intensive to manufacture, but they have higher emissions from day-to-day use.

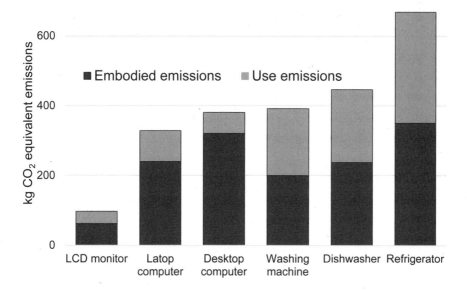

Figure 11.4. Estimated lifetime emissions for several different appliance types. Based on data in Gonzalez, A., Chase, A., and Horowitz, N., "What We Don't Know About Embodied Energy and Greenhouse Gases for Electronics, Appliances and Light Bulbs," American Council for an Energy-Efficient Economy, Summer Study on Energy Efficiency in Buildings, 2012, pp. 140–50. The equivalent CO_2 emissions are based on both manufacturing and use in a region where the majority of electricity is derived from combustion of natural gas.

The message from figure 11.4 is that if you want to reduce your GHG emissions, don't replace your computer or phone frequently. If you want to reduce your emissions from the washing machine or the dishwasher, use them only sparingly.

Reducing Methane Emissions

Most of our methane emissions come from producing food (51%), from extracting and processing fossil fuels (29%), and from landfills and sewage (20%).[12]

Food

In order to control our food-related emissions, we obviously need to look carefully at what we eat. Figure 11.5 provides an estimate of the GHG emissions from various foods; it includes emissions attributed

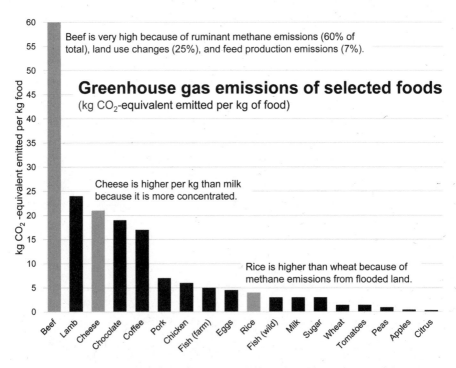

Figure 11.5. Greenhouse gas emissions attributed to the production of various food types. Based on a diagram by Hannah Ritchie, "Food Choice vs Eating Local," *Our World in Data*, January 24, 2020.

to land-use changes, food production itself (e.g., machinery, chemicals), feed for animals, processing, transportation, packaging, and retail sales. Although the scale is expressed in CO_2-equivalent, a majority of the GHG emissions from most of these foods are in the form of methane.

At 60 kg of CO_2-equivalent per kg of food produced, beef is far and away the biggest contributor to GHG emissions. Most of that (60%) is methane produced in the rumen of the cows, 25% comes from the conversion of forest to pasture (because of the lower uptake of CO_2 from grass than from trees), 7% is from the production of grain and other feeds, 5% is from processing, and 4% is from transportation and packaging.

Although sheep are also ruminants, lamb and mutton are not as GHG-intensive as beef because their land-use implications are much lower, and sheep aren't as gassy as cows. But the best options for carnivores are pork, chicken, and fish, which have about one-tenth of the GHG emissions of beef.

Coffee and chocolate are quite big GHG contributors per kilogram, but most of us don't consume a lot of either of these in typical day. An average cup of coffee is made with about 7 g of beans, and you would have to drink 480 of those cups a day to produce as much CO_2 as from eating a typical (100 g) serving of beef.

Cheese is high on the GHG list because most of it comes from cows, and it is much higher than milk because it takes roughly 10 kg of milk to make 1 kg of cheese.

Rice is higher on the list than wheat because it is mostly grown in flooded fields that emit methane. That said, a typical serving of rice results in approximately 4% of the emissions of a typical serving of beef.

Fugitive Emissions from Fossil Fuels

Production of fossil fuels results in massive amounts of waste gas (mostly methane) at many points within the process. Methane is released from coal deposits as they are mined; it is produced along with oil at oil wells (and may or may not be captured or flared[13]); and it is inadvertently or carelessly released from natural gas wells. Methane is also released by leakage from refineries, pipelines, and machines that burn oil and gas. The amount of leakage is just a few percent of the amount of fossil-fuel production,[14] but that represents a huge amount of methane. It's a serious problem because methane is 25 to 30 times as potent a greenhouse gas as the carbon dioxide that would be produced if the methane were captured and used.

The best way for us to reduce fugitive emissions is to dramatically reduce our consumption of fossil fuels, but there are some steps that must be taken in the time that it's going to take us to wrestle our fossil-fuel use down to nearly zero. For the most part, these are not things that we can do as individuals, but we can put pressure on governments to tighten fugitive gas regulations and on fossil fuel corporations to work harder on detecting leaks, repairing leaking equipment promptly, capturing gas that would otherwise be lost or flared, and either reinjecting it into the reservoir, using it to generate electricity, or transporting it to somewhere that it can be used.

Landfills and Sewage

As the waste in landfills breaks down, it emits a range of by-products, including methane and carbon dioxide. Those gases migrate toward the surface, and unless there are measures in place to prevent it, they escape to the atmosphere. Many modern landfills have impermeable covers over sections that are no longer being filled and wells to facilitate extraction of the gases. These gases are either flared to convert the methane to carbon dioxide, or the methane is diverted to a facility where it can be used to generate heat or electricity.

Methane is also produced in sewage treatment plants, and in that situation, it can be relatively easily captured and used to generate heat and/or electricity to run the plant.

Although methane from landfills can be captured and used, much of it is lost, even in the best-designed operations, and at many landfills nothing is done to prevent its release. All of us can help to reduce the methane emission problem by reducing our waste stream, by recycling as much as is possible in our jurisdiction, and especially, by helping to divert the organic wastes that produce methane in land-fills. This can be achieved by wasting less food (much less!), composting at home (which can be done even in apartment buildings), and taking advantage of the growing efforts of municipalities to collect organic wastes and deal with them in large-scale composting facilities.

In summary, we all have a role to play in significantly reducing our greenhouse gas emissions. Our number one goal needs to be to drive much less and walk and bike, use transit, and plan our trips more carefully. If we must drive, then it should be in an electric vehicle.

Only a small fraction of us fly frequently, but those who do have a significant climate impact. A recent analysis of air travel has shown that 89% of the population didn't fly anywhere in 2018 and that 1% of the global population is responsible for more than 50% of air-travel emissions.[15] The fact that many of us living in wealthy nations don't fly very much (many not all) makes it abundantly clear that, for the vast majority of people, there is really no need to be flying all over the place. It is an indulgence that we don't need and cannot afford.

Buildings use a lot of energy, partly because many of them are too big, are not energy efficient, and don't have energy-efficient or climate-friendly heating and cooling systems. Changing that will be slow, but in the meantime, we can all help by minimizing our heating and cooling demands.

Finally, what we eat can have significant climate implications. Most of us need to eat less meat, especially less beef.

It's clear that we all need to make some changes to the way we do things, and we need to stop waiting for some crisis or signal or incentive or law to get us started. The climate crisis is here now, and it is becoming increasingly obvious with every year that passes. It is negatively affecting billions of people and is putting ecosystems everywhere into stress. The personal costs of making changes are real, but they are dwarfed by the societal costs of not making changes. That imbalance will continue to get worse if we continue to delay. The dollar costs of failing to make changes are already measured in numbers that we have difficulty comprehending, and those are going to rise exponentially if we don't act. As my friend Lynne Quarmby writes,

"There is a yawning chasm of difference between how bad things will get if we continue business-as-usual and how bad they will be if we get off fossil fuels as soon as possible."[16]

The Climate Impact of COVID-19

As I write this late in 2020, many countries are in the throes of a second wave of the COVID-19 pandemic. It is much more severe than the first time around, and since the case numbers are still trending upward in most countries, it appears likely to get even worse before it gets better. Health, social, business, democratic, and educational systems are seriously hobbled; hotels are mostly empty; nearly all cruise ships are tied up; and tens of thousands of commercial aircraft are grounded.

While these are difficult times for many people, there may be an upside to the COVID-related restrictions from the perspective of

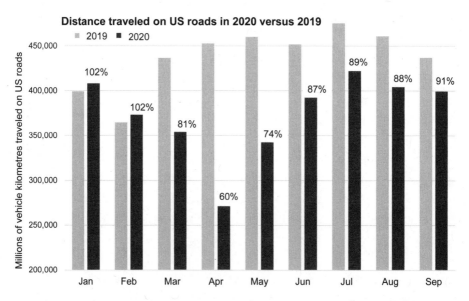

Figure 11.6. Monthly average distance traveled on US roads in the first nine months of 2020 and 2019. The percentages are for 2020 versus 2019. Based on data from the Federal Highway Administration of the U.S. Department of Transportation, fhwa.dot.gov/policyinformation/travel_monitoring/tvt.cfm, October 2020.

climate change. Many office employees are working from home (although, sadly, many people are not working at all), and many schools and universities have gone online, so road traffic is down, especially in countries where transportation is dominated by private cars, like the US and Canada.

The total distance traveled on US roads is monitored by the Federal Highway Administration of the U.S. Department of Transportation, and the numbers for 2020 and 2019 are shown on figure 11.6. The roads were as busy as normal in January and February of 2020, but traffic dropped significantly in March, April, and May compared with the same months in 2019, with the lowest traffic in April at 60% of "normal." Traffic recovered in June, July, August, and September to approximately 90% of normal levels. From March to September 2020, there was an average 80% decrease in traffic on US roads compared with 2019.

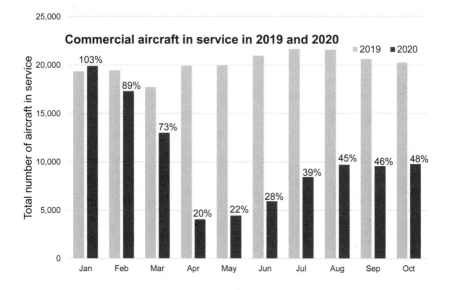

Figure 11.7. Total number of aircraft (passenger and freight) in service for all world airlines in the first ten months of 2019 and 2020. The percentages are for 2020 versus 2019. Based on International Civil Aviation Organization data, data.icao.int/coVID-19/aircraft.htm, accessed October 2020.

Demand for air travel dropped dramatically in 2020, largely because of COVID-related international travel restrictions and national policies discouraging business, social, and family gatherings. One measure of the decrease in air traffic comes from the number of aircraft in service as reported by the International Civil Aviation Organization. In January 2020, there were slightly more aircraft in service than in January 2019, but by April that number was down to 20% of the previous year (figure 11.7). The aviation industry recovered slightly over the spring and summer. On average, there were 46% as many aircraft in service in February through October 2020 compared with the same period in 2019.

In a study published in October 2020, Liu and coauthors estimated CO_2 emissions from various types of activities during the first half of 2020.[17] They document decreases in emissions related to road and air travel that are similar to those described above but show that emission decreases in other areas were smaller: about 5% in electricity generation and industry and only 2% in buildings. They estimate that global CO_2 emissions averaged 8.8% below 2019 levels during the first six months of 2020, with the lowest levels in April (figure 11.8). A re-

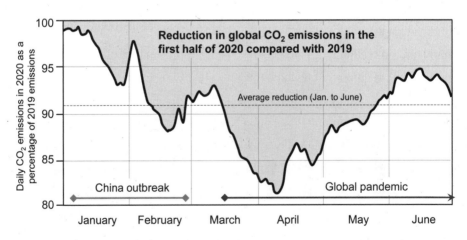

Figure 11.8. Estimated total CO_2 emission reductions during the COVID pandemic in the first half of 2020. Based on data in Liu et al., "Near-real-time Monitoring of Global CO_2 Emissions Reveals the Effects of the COVID-19 Pandemic," *Nature Communications*, V. 11, 5172, 2020.

port published in mid-December 2020 provides an update that there has been an overall 7% decrease in CO_2 emissions in 2020 compared with 2019,[18] the largest year-over-year decrease ever recorded.

Of course, the more important consideration is whether the observed drop in CO_2-producing activity and the estimated drop in emissions has had any impact on the levels of CO_2 in the atmosphere. In fact, it's too early to say. The sharpest emission decline in the past 50 years was during the recession of the early 1980s: an overall 11% drop in CO_2 emissions from 1980 to 1983. This downturn did result in a slight reduction in the rate of increase of CO_2 in the atmosphere, but that didn't happen until late 1983 and into the middle of 1984.

The Mauna Loa CO_2 curve for 2017 to 2020 is shown on figure 11.9. As of October 2020, there was no evidence of an inflection that could be related to COVID-19, although, based on experience from the 1980s recession, we wouldn't expect to see that yet. If the COVID downturn

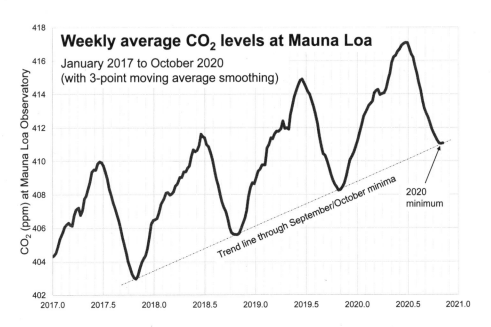

Figure 11.9. CO_2 levels at Mauna Loa from January 2017 to October 2020. Based on data from Scripps Institute of Oceanography, U. of California, San Diego, scrippsco2.ucsd.edu/data/atmospheric_co2/primary_mlo_co2_record .html

in travel continues for another year or more, then we might see a small reduction in the rate of CO_2 increase, but this is not the silver bullet that is going to slay climate change. Instead, it shows us that life can be bearable without a lot of discretionary travel, and that we need to take much more ambitious steps in that direction.

Of greater significance than a small and likely temporary reduction in emissions, COVID-19 has shown us that we do have the capacity to mobilize vast resources and the will to work together to get through a crisis. Governments have found the funds to help those forced out of work; hospitals and medical staff have been truly heroic; pharmaceutical scientists have worked at warp speed to design, test, and manufacture vaccines; and the majority of individuals have shown that they can modify their behaviors for the collective good.

If we can do this to fight COVID, surely we can bring the same kind of resolve and effort into the fight against climate change, which poses a much greater risk to our existence here on Earth.

Notes

Preface

1. Originally published in *Poem-a-Day* on January 20, 2017, by the Academy of American Poets: poets.org/poem/let-them-not-say.

Chapter 1: What Controls the Earth's Climate?

1. If you want to get technical, when talking about gas concentrations in ppm, we usually mean ppmv, or "ppm by volume." So, for a gas at a concentration of 1 ppmv, there is 1 cubic cm of that gas in every 1 million cubic cm (1 cubic m) of air.
2. In case you're wondering, the warming of Earth's surfaces is almost exclusively the work of the sun. Some heat comes from the Earth's interior, but that contribution is tiny, approximately 0.03% of the amount that we get from the sun.
3. Ozone and the chlorofluorocarbons are complicated, and there is a lot of potential for confusion around them. Ozone that is present in the lower atmosphere (called ground-level ozone) is a GHG, while ozone in the stratosphere is important in preventing ultraviolet radiation from reaching the surface. CFCs in the lower atmosphere also behave as GHGs, while CFCs in the stratosphere are responsible for breaking down ozone, putting us at risk of too much UV radiation. Since the enactment of the Montreal Protocol in 1987, the "hole in the ozone" problem in the stratosphere has been slowly decreasing, but that doesn't have climate implications. Down here in the troposphere, the ozone and CFCs are still important as GHGs.
4. Klimont, Z., et al., 2017, "Global Anthropogenic Emissions of Particulate Matter Including Black Carbon," *Atmos. Chemistry and Physics*, V. 17, pp. 8681–723. They estimate that the rate for 2010 was 7.2 million tonnes of black carbon, up from 6.6 million tonnes in 2000.
5. The arguments included here are selected from lists compiled by *Skeptical Science*, *CBS News*, *Wikipedia*, California Governor's Office of Planning and Research, and *ScienceAlert*.

Chapter 2: A Slowly Warming Sun

1. Oliver, Mary, *New and Selected Poems, Volume 1*, Beacon Press, Boston, 1992.
2. Short-term solar variations are discussed in chapter 7.
3. The history of the sun (and other stars) is summarized by David Taylor of Northwestern University at faculty.wcas.northwestern.edu/~infocom/The%20Website/index.html.
4. Luminosity is a measure of the amount of energy emitted by the sun, and that is directly proportional to how much solar energy is received on the surface of the Earth.
5. Virtually all references to "temperature" in this volume refer to the Earth's mean annual temperature (MAT), the average temperature over every square kilometer of the Earth, including the tropics and the poles, the oceans, and the land—and also averaged over an entire year.
6. There are no fossils as old as 4 Ga, but there is chemical evidence of life in rocks that old, specifically carbon deposits that have the isotopic signature of being formed by living organisms. Tashiro, T., et al., 2017, "Early Trace of Life from 3.95 Ga Sedimentary Rocks in Labrador, Canada," *Nature*, V. 549, pp. 516–18. The oldest undisputed evidence of life is in the form of fossils dating to about 3.5 Ga.
7. The geological time scale is described in chapter 1.
8. Becker, S., et al., "Unified Prebiotically Plausible Synthesis of Pyrimidine and Purine RNA Ribonucleotides," *Science*, V. 366, pp. 76–82, 2019.
9. Blankenship, R., "Early Evolution of Photosynthesis," *Plant Physiology*, V. 154, pp. 434–8, 2010.
10. The reaction goes like this: $CH_4 + 2O_2 \rightarrow CO_2 + 2H_2O$. One molecule of methane reacts with two oxygens to produce one carbon dioxide and two molecules of water. In other words, methane is being oxidized to carbon dioxide. Of course, carbon dioxide is also a greenhouse gas, but it is only about 1/20 as potent a warming agent as methane.
11. Evidence from ancient rocks suggests that the Huronian glaciation lasted for at least 40 million years (from 2.29 to 2.25 Ga). It is thought to have been widespread, based on glacial evidence in rocks of that time from Canada, US, Europe, South Africa, India, Australia, and Brazil and may have affected low latitudes as well as polar regions. Tang, H., and Chen, Y., "Global Glaciations and Atmospheric Change at ca. 2.3 Ga," *Geoscience Frontiers*, V. 4 (5), pp. 583–96, 2013. The oceans may have been mostly frozen over.
12. Lovelock, J., "Gaia as Seen Through the Atmosphere, *Atmospheric Environment*," V. 6, pp. 579–80, 1972.

13. Lovelock, J., and Margulis, L., "Atmospheric Homeostasis by and for the Biosphere: The Gaia Hypothesis," *Tellus*, V. 26, pp. 2–10, 1974.

14. Lovelock, J., *Gaia: A New Look at Life on Earth*, Oxford University Press, 1979.

15. Watson, A., and Lovelock, J., "Biological Homeostasis of the Global Environment: The Parable of Daisyworld. *Tellus*," V. 35B, pp. 284–89, 1983. As anyone with a search engine can discover, a great deal has been written about Daisyworld since 1983.

16. Limestone consists mostly of the mineral calcite, $CaCO_3$.

17. Graphite is pure carbon. It cannot be used as a fuel.

18. Wolf, E., and Toon, O., "Delayed Onset of Runaway and Moist Greenhouse Climates for Earth," *Geophys. Res. Lett.*, V. 41, pp. 167–72, 2014.

Chapter 3: Sliding Plates and Colliding Continents

1. Wegener, A., *Die Entstehung der Kontinente und Ozeane*, 4. Auflage, Friedrich Vieweg & Sohn, Braunschweig, 1929 (*The Origin of Continents and Oceans*, 4th ed., John Biram, trans., Dover Publications, Mineola, NY, 1966). Wegener conceived the idea of continental drift in 1910, but his theory wasn't widely accepted by the geological community until the mid-1960s, 35 years after his death. It is now fundamental to our understanding of the Earth and its processes, and it has important implications for climate change. For more on plate tectonics, see opentextbc.ca/physi calgeology2ed/part/chapter-10-plate-tectonics.

2. Albedo is a measure of the reflectivity of the Earth's surfaces. It is discussed in chapter 1.

3. Plants first came onto land at around 450 Ma.

4. The Cryogenian Period lasted from 720 to 635 Ma and included two snowball glaciations: the Sturtian, from about 717 to 660 Ma, and the Marinoan, 650 to 635 Ma.

5. Piper, J., "Dominant Lid Tectonics Behaviour of Continental Lithosphere in Precambrian Times: Palaeomagnetism Confirms Prolonged Quasi-Integrity and Absence of Supercontinent Cycles," *Geoscience Frontiers*, V. 9, pp. 61–89, 2018.

6. Crowley, T., Hyde, W., and Peltier, W., "CO_2 Levels Required for Deglaciation of a 'Near-snowball' Earth," *Geophys. Res. Lett.*, V. 28, pp. 283–6, 2001.

7. Hoffmann et al., "Snowball Earth Climate Dynamics and Cryogenian Geology-geobiology," *Science Advances*, V. 3, pp. 1–43, 2017.

8. There are 131 mountains in the world that are over 7,000 m tall. All of them, yes all of them, are part of the Himalayas or adjacent ranges. Most of the mountains taller than 5,000 m are also in the Himalayan region.

9. Hydrolysis is the process through which a molecule is split apart by water. In the context of mineral weathering, it can be represented like this: $CaAl_2Si_2O_8 + H_2O + CO_2 + \frac{1}{2}O_2 \rightarrow Al_2Si_2O_5(OH)_4 + (Ca^{2+} + CO_3^{2-})$ in which feldspar $(CaAl_2Si_2O_8)$ reacts with water, carbon dioxide, and oxygen to form the clay mineral kaolinite $(Al_2Si_2O_5(OH)_4)$ along with calcium and carbonate ions in solution. The key thing happening here is that carbon dioxide is coming out of the atmosphere. It will eventually get fixed into a mineral-like calcite $(CaCO_3)$.

10. Bartoli, G., et al., "Final Closure of Panama and the Onset of Northern Hemisphere Glaciation," *Earth and Planetary Science Letters*, V. 237, pp. 33–44, 2005.

Chapter 4: Cooling and Warming from Volcanic Eruptions

1. Richerus of Sens, "Gesta Senoniensis Ecclesiae," in *Societas Aperiendis Fontibus Rerum Germanicarum Medii Aevi* (ed.), 1267: *Monumenta Germaniae Historica*, Scriptores 25, Hahn's, Hannover, pp. 333–4, 1880. Richerus (1218–1267), a Benedictine monk of Senones (Sens, France), recorded the cold, wet, and dull conditions during the non-summer of 1258. The source of the obscured sky and unseasonable weather was the massive eruption of the Samalas volcano, on Lombok, Indonesia, in 1257.

2. A distance of 2,240 leagues is 12,445 km, which is equivalent to about one-third of the way around the world.

3. An example of a wet mineral is serpentine, $(Mg,Fe)_3Si_2O_5(OH)_4$. On sufficient heating, serpentine will be converted to olivine and pyroxene, and water will be released.

4. CO_2 is removed from the atmosphere through uptake by plants, but when plants die and decay, that carbon returns to the atmosphere. It is also removed through dissolution into the ocean, but the surface of the ocean can take up only so much CO_2 before becoming saturated, and so the very slow rate (centuries to millennia) of ocean turnover becomes the limiting factor. See section 2.10 in the IPCC Fourth Assessment Report, ipcc.ch/site/assets/uploads/2018/02/ar4-wg1-chapter2-1.pdf.

5. Most of the information about the 1873 Laki eruption is based on Thordarson, T., and Self, S., "Atmospheric and Environmental Effects of the 1783–1784 Laki Eruption: A Review and Reassessment," *J. Geophys. Res.*, V. 108 (D1), 4011, 2003.

6. Neale, G., "How a Volcano in Iceland Helped Spark the French Revolution," *Guardian*, April, 15, 2010.

7. By the author, from figure 10 in Thordarson, T., and Self, S., "Atmospheric and Environmental Effects of the 1783–1784 Laki Eruption: A Review and Reassessment," *J. Geophys. Res.*, V. 108 (D1), 4011, 2003.

8. Most of the information about the 1257 Samalas eruption is based on Vidal, C., et al., "The 1257 Samalas Eruption (Lombok, Indonesia): The Single Greatest Stratospheric Gas Release of the Common Era," *Science Reports*, V. 6, 34868, 2016.

9. Robock, A., et al., "Did the Toba Volcanic Eruption of ~74k BP Produce Widespread Glaciation?" *Journal of Geophysical Research*, V. 114 (D10), D10107, 2009.

10. Svensson, A., et al., "Direct Linking of Greenland and Antarctic Ice Cores at the Toba Eruption (74 ka BP)," *Climates Past*, V. 9, pp. 749–66, 2013.

11. Kennett, J., et al., "Santa Barbara Basin Sediment Record of Volcanic Winters Triggered by Two Yellowstone Supervolcano Eruptions at 639 ka," Geol. Soc. of Amer. Ann. Mtg., Seattle, 2017, Paper no. 393-7.

12. Dessert, C., et al., "Erosion of Deccan Traps Determined by River Geochemistry: Impact on the Global Climate and the $^{87}Sr/^{86}Sr$ Ratio of Seawater," *Earth and Planetary Science Letters*, V. 188, pp. 459–74, 2001. By way of comparison, the current mass of CO_2 in the atmosphere is about 75 trillion tonnes. The 35,000-year interval shown here represents a major pulse of volcanism at that time, although it's evident that the entire Deccan eruption cycle had started long before this event, and had a duration approaching 1 million years.

13. Richards, M., et al., "Triggering of the Largest Deccan Eruptions by the Chicxulub Impact," *Geol. Soc. Amer. Bulletin*, V. 127 (11–12), pp. 1507–20, 2015.

14. Burgess, S., and Bowring, S., "High-precision Geochronology Confirms Voluminous Magmatism Before, During, and After the Earth's Most Severe Extinction," *Science Advances*, V. 1, no. 7, 2015.

15. Prior to the "information age," many large eruptions were simply undetected, and in many parts of the world, those that were detected were not recorded.

Chapter 5: Earth's Orbital Variations

1. Milutin Milanković (pronounced Milan-ko-vitch), born in 1879 in what was is now Croatia, was an engineer, physicist, and mathematician. He calculated the differences in insolation levels at different latitudes on Earth that result from variations in the Earth's orbital parameters and theorized that these differences controlled the growth and decline of glaciers during the Pleistocene. His theory was widely rejected until 18 years after his death in 1958. From "Milutin Milanković," famous scientists.org, March 31, 2018.

2. The Earth is not unique in having an elliptical orbit. All planetary bodies have elliptical orbits, and the orbits of some, such as comets, are much

more elliptical than the Earth's. If you know something about ellipses, you'll know that they have two foci. In the sun-Earth system (as in all other planetary systems), the sun is located at one of the foci, and the other one is empty. All orbiting objects also have tilted axes. Some, Uranus for example, have much higher tilts than the Earth.

3. Latitude 65°, north or south, is ideal for the growth of glaciers because summers can be cool enough for some of the winter snow to last through the year and because there is generally more snow than in areas to the north or south.

4. Hays, J., Imbrie, J., and Shackleton, N., "Variations in the Earth's Orbit: Pacemaker of the Ice Ages," *Science*, V. 194, pp. 1121–32, 1976.

5. Time calibration of ice cores is done by counting layers in the ice and also by radiometrically dating volcanic ash layers in the ice and correlating ash layers with eruptions for which the dates are known.

6. For example, see Berger, A., and Loutre, M-F., "Climate: An Exceptionally Long Interglacial Ahead?" *Science*, V. 297, pp. 1287–8, 2002.

Chapter 6: Moving Heat with Ocean Currents

1. These words of historian Antonio de Herrera y Tordesillas describe a log entry Ponce de Leon made while sailing along the eastern edge of Florida in 1513. Herrera recorded them in 1615, in his *Historia general de los hechos de los castellanos en las Islas y Tierra Firme del mar Océano que llaman Indias Occidentales* (*General History of the Deeds of the Castilians on the Islands and Land of the Ocean Sea Known as the West Indies*), ch. X.

2. Gyory, J., Mariano, A., and Ryan, E., "The Gulf Stream," Ocean Surface Currents, oceancurrents.rsmas.miami.edu/atlantic/gulf-stream.html. Retrieved April 5, 2020.

3. Dai, A., and Trenberth, K., "Estimates of Freshwater Discharge from Continents: Latitudinal and Seasonal Variation," *J. of Hydrometeorology*, V. 3, pp. 660–87, 2002.

4. Yes, liquid water can exist below 0°C, especially if it is salty.

5. See Ackerman, Steven and Knox, John A., *Meteorology: Understanding the Atmosphere*, ch. 14, Brooks Cole, 2002.

6. Robson, J., et al., "Atlantic Overturning in Decline?" *Nature Geosci.*, V. 7, pp. 2–3, 2014.

7. Based on data from the NASA Goddard Institute for Space Studies, data.giss.nasa.gov/gistemp/tabledata_v4/GLB.Ts+dSST.txt.

8. Barker, S., et al., "Interhemispheric Atlantic Seesaw Response During the Last Deglaciation," *Nature*, V. 457, pp. 1097–102, 2009.

9. The Niño 3.4 region is between 5° north and 5° south, and between

120° and 170° west. The Niño 3.4 index is determined by comparing the surface water temperature within that region for any 3-month period with the average for a 30-year period (1986 to 2015). A map showing the extent of that region is at ncdc.noaa.gov/teleconnections/enso/indicators/sst/

10. Based on data from the Japan Meteorological Agency, data.jma.go.jp/gmd/kaiyou/english/long_term_sst_global/glb_warm_e.html.

11. Freund, M., et al., "Higher Frequency of Central Pacific El Nino Events in Recent Decades Relative to Past Centuries," *Nature Geoscience*, V. 12, p. 450, 2019.

12. Wang, B., et al., "Historical Change of El Niño Properties Sheds Light on Future Changes of Extreme El Niño," *Proc. Natl. Acad. of Science*, V. 116, pp. 22512–517, 2019.

Chapter 7: Short-term Solar Variations

1. These are descriptions based on naked-eye observations of the sun by Chinese astronomers in the second and third centuries AD, as translated from imperial court records by Yau, K., and Stephenson, F., "A Revised Catalogue of Far-eastern Observations of Sunspots (165 BC to AD 1918)," *Q. Jour. Royal Astronomical Society*, V. 29, pp. 175–97, 1988.

2. Fabricii, J., *Phrysii De Maculis in Sole observatis, et apparente earum cum Sole conversione, Narratio*, etc., Witebergae,1611. The telescope has been invented only a few years earlier (1608) by Hans Lippershey, in Middelburg, Zeeland (now part of the Netherlands).

3. Galilei, G., *Istoria e Dimostrazioni Intorno Alle Macchie Solari e Loro Accidenti Rome* (*History and Demonstrations Concerning Sunspots and Their Properties*), 1613. An animation of Galileo's drawings made in 1613 is available at academo.org/demos/galileos-sunspots.

4. In the early days of a sunspot's life, its proper motion (actual motion across the sun's surface) can be in the order of 2° per day (or about 25,000 km/day), but the rate drops to a fraction of that over its life. Large sunspots tend to move faster than small ones. See, for example, Ambastha, A., and Bhatnagar, A., "Sunspot Proper Motions in Active Region NOAA 2372 and Its Flare Activity During SMY Period of 1980 April 4–13," *J. Astrophys. & Astron.*, V. 9, pp. 137–54, 1988.

5. Stefani, F., Giesecke, A., and Weier, T., "A Model of a Tidally Synchronized Solar Dynamo," *Solar Physics*, V. 294, Article no. 60, 2019.

6. Arlt, R., and Vaquero, J., "Historical Sunspot Records," *Living Rev. in Solar Phys.*, V. 17, 2020.

7. As measured in space at the distance of the Earth's orbit from the sun.

8. Wagner, S., and Zorita, E., "The Influence of Volcanic, Solar and CO_2 Forcing on the Temperatures in the Dalton Minimum (1790–1830): A Model Study," *Climate Dynamics*, V. 25, 2005.

9. Eddy, J., "The Maunder Minimum," *Science*, V. 192, pp. 1189–202, 1976.

10. Rigozo, N., et al., "Reconstruction of Wolf Sunspot Numbers on the Basis of Spectral Characteristics and Estimates of Associated Radio Flux and Solar Wind Parameters for the Last Millennium," *Solar Physics*, V. 203, pp. 179–91, 2001.

11. Fagan, B., *The Little Ice Age: How Climate Made History, 1300–1850*, Basic Books, 2001.

12. Li, Y., Lu, X., and Li, Y., "A Review on the Little Ice Age and Factors to Glacier Changes in the Tian Shan, Central Asia," in *Glacier Evolution in a Changing World*, D. Godone (ed.), IntechOpen, 2017.

13. Luckman, B., Masiokas, M., and Nicolussi, K., "Neoglacial History of Robson Glacier, British Columbia," *Canadian Journal of Earth Sciences*, V. 54, pp. 1153–64, 2017.

14. Luckman, B., "Calendar-dated, Early Little Ice Age Glacier Advance at Robson Glacier, British Columbia, Canada," *The Holocene*, V. 5, pp. 149–59, 1995.

15. Solomina, O., et al., "Holocene Glacier Fluctuations," *Quaternary Science Reviews*, V. 111, pp. 9–34, 2015.

Chapter 8: Catastrophic Collisions

1. This quotation, in reference to the extraterrestrial object (a comet or a meteorite, it doesn't really matter) that struck the Earth 66 million years ago, precisely at the end of the Cretaceous Period and the start of the Paleogene Period (the K-Pg boundary), is from Alvarez, W., "*T. rex*" *and the Crater of Doom*, Princeton University Press, 2013. The term that might be more familiar is "K-T boundary," where K is for Cretaceous and T for Tertiary. K-Pg is more technically correct because both Cretaceous and Paleogene are period names, while Tertiary is an outdated name for the Paleogene and the Neogene combined. In 1980, Walter Alvarez, Luis Alvarez (his father), Frank Asaro, and Helen Michel, were the first to suggest that the massive extinction at the end of the Cretaceous was the result of an extraterrestrial impact: "Extraterrestrial Cause for the Cretaceous:Tertiary Extinction," *Science*, V. 208, pp. 1095–108, 1980.

2. Robertson, D., et al., "Survival in the First Hours of the Cenozoic," *Geological Society of America Bulletin*, V. 116, 2004.

3. Many of the following details are from Gulick, S., et al., "The First Day of the Cenozoic," *Proc. Natl. Acad. Sci.*, V. 116, no. 39, 2019.

4. Kornei, K., "Huge Global Tsunami Followed Dinosaur-Killing Asteroid

Impact," *Eos*, V. 99, 2018, a summary of the research by Range, M., et al., "The Chicxulub Impact Produced a Powerful Global Tsunami," presentation at the Amer. Geophysical Union Fall meeting, Washington, DC, December 2018.

5. Bardeen, C., et al., "On Transient Climate Change at the Cretaceous-Paleogene Boundary Due to Atmospheric Soot Injections," *Proc. Natl. Acad. Sci.*, V. 114, 2017, E7415–24.

6. MacLeod, K., et al., "Postimpact Earliest Paleogene Warming Shown by Fish Debris Oxygen Isotopes (El Kef, Tunisia)," *Science*, V. 360, pp. 1467–69, 2018.

7. "Cretaceous-Paleogene extinction event," *Wikipedia*.

8. Bottke, F., and Norman, M., "The Late Heavy Bombardment," *Ann. Rev. Earth Planet. Sci.*, V. 45, pp. 619–47, 2017.

9. Hildebrand, A., et al., "Chicxulub Crater: A Possible Cretaceous/Tertiary Boundary Impact Crater on the Yucatan Peninsula, Mexico," *Geology*, V. 19, pp. 867–71, 1991.

10. Smitz, B., et al., "An Extraterrestrial Trigger for the Mid-Ordovician Ice Age: Dust from the Breakup of the L-chondrite Parent Body," *Science Advances*, V. 5, eaax4184, 2019. The authors suggest that meteoritic dust blocking sunlight was the main reason for cooling but that the enhanced input of iron into the ocean may also have fertilized algal growth, reducing the CO_2 content of the atmosphere.

11. Wielicki, M., Harrison, M., and Stockli D., "Popigai Impact and the Eocene/Oligocene Boundary Mass Extinction," Goldschmidt conference presentation, Sacramento, CA, 2014.

12. Zolensky, M., et al., "Flux of Extraterrestrial Materials," in D. Lauretta and H. McSween, eds., *Meteorites and the Early Solar System II*, Tucson: University of Arizona Press, pp. 869–88, 2006.

13. The Center for Near Earth Object Studies, Jet Propulsion Laboratory, California Institute of Technology, NASA, cneos.jpl.nasa.gov/about/search_program.html

Chapter 9: A Plague of Humans

1. From the film *The 11th Hour*, Leila Connors, Director, 2007.

2. Zhu, Z., et al., "Hominin Occupation of the Chinese Loess Plateau Since About 2.1 Million Years Ago," *Nature*, V. 559, pp. 608–12, 2018.

3. Ibid.; and Bae, C., Douka, K., and Petraglia, M., "On the Origin of Modern Humans: Asian Perspectives," *Science*, V. 358 (6368), 2017.

4. Insolation curve based on data from Berger, A., and Loutre, M-F., "Insolation Values for the Climate of the Last 10 Million Years." *Quaternary*

Science Reviews, V. 10, pp. 297–317 (Supplement: Parameters of the Earth's orbit for the last 5 Million years in 1 kyr resolution), 1991.

5. Zeder, M., "The Origins of Agriculture in the Near East," *Current Anthropology,* V. 52, Supplement 4, pp. S221–35, 2011.

6. Zuo, X., et al., "Dating Rice Remains Through Phytolith Carbon-14 Study Reveals Domestication at the Beginning of the Holocene," *Proc. Nat. Acad. of Sci.,* V. 114, pp. 6486–91, 2017.

7. Piperno, D., et al., "Starch Grain and Phytolith Evidence for Early Ninth Millennium B.P. Maize from the Central Balsas River Valley, Mexico," *Proc. Nat. Acad. of Sci.,* V. 106, pp. 5019–24, 2009.

8. Blaustein, R., "William Ruddiman and the Ruddiman Hypothesis," *Minding Nature,* V. 8, no.1, 2015, humansandnature.org, accessed September 2020.

9. Ruddiman, W., and Thomson, J., "The Case for Human Causes of Increased Atmospheric CH_4 Over the Last 5000 Years," *Quat. Sci. Rev.,* V. 20, pp. 1769–77, 2001; Ruddiman, W., "The Anthropogenic Greenhouse Era Began Thousands of Years Ago," *Clim. Change,* V. 61, pp. 261–93, 2003.

10. Ruddiman, W., "The Early Anthropogenic Hypothesis: Challenges and Responses," *Rev. Geophys.,* V. 45, RG4001, 2007.

11. McEvedy, C., and Jones, R., *Atlas of World Population History, Facts on File,* New York, pp. 342–51, 1978; "World Population Growth," *Our World in Data,* accessed September 2020.

12. Pirani, S., *Burning Up: A Global History of Fossil Fuel Consumption,* London, Pluto Press, 2018.

13. Data from the International Energy Agency, iea.org/commentaries/iea-releases-new-edition-of-global-historical-data-series-for-all-fuels-all-sectors-and-energy-balances, accessed September 2020. Most of the remaining amount was generated by hydro and nuclear.

14. Kammen D., et al., *IPCC Special Report on Renewable Energy Sources and Climate Change Mitigation.* Prepared by Working Group III of the Intergovernmental Panel on Climate Change, Cambridge University Press, 2011.

Chapter 10: Tipping Points

1. Hansen, J. E., "Is There Still Time to Avoid 'Dangerous Anthropogenic Interference' with Global Climate?" presentation at the American Geophysical Union, San Francisco, December 6, 2005.

2. Canadian journalist Gwynne Dyer, "Warming Accelerates in Unprecedented Way," *Otago Daily Times,* February 5, 2008.

3. Seba, T., "Clean Disruption of Energy and Transportation," lecture pre-

sented at Clean Energy Action, Boulder, CO, 2017, youtube.com/watch ?v=2b3ttqYDwF0, accessed September 2020.

4. Based on data from the California Department of Forestry and Fire Protection: 17,000 km² is about the size of Connecticut plus Delaware, or three times the area of Prince Edward Island, fire.ca.gov/stats-events.

5. "2020 Western United States wildfires," *Wikipedia*.

6. Garfin, G., et al., "Southwest: The Third National Climate Assessment," in J. M. Melillo, T. C. Richmond, and G. W. Yohe, eds., *Climate Change Impacts in the United States: The Third National Climate Assessment*, pp. 462–86, U.S. Global Change Research Program.

7. Goss, M., et al., "Climate Change Is Increasing the Likelihood of Extreme Autumn Wildfire Conditions Across California," *Environ. Res. Lett.*, V. 15, 2020.

8. Swain, D., et al., "Increasing Precipitation Volatility in Twenty-First-Century California," *Nature Climate Change*, V. 8, pp. 427–33, 2018.

9. Arneth, A., et al., *IPCC Special Report on Climate Change, Desertification, Land Degradation, Sustainable Land Management, Food Security, and Greenhouse Gas Fluxes in Terrestrial Ecosystems Summary for Policymakers*, 2019, ipcc.ch/site/assets/uploads/2019/08/Fullreport.pdf, accessed September 2020.

10. Stevens-Rumann, C., & Morgan, P., "Tree Regeneration Following Wildfires in the Western US: A Review," *Fire Ecology*, V. 15, 2019.

11. Davis, K., et al., "Wildfires and Climate Change Push Low-elevation Forests Across a Critical Climate Threshold for Tree Regeneration," *Proceedings of the National Academy of Sciences*, V. 116, pp. 6193–8, 2019.

12. Zachos, J., et al., "Paleocene-Eocene Thermal Maximum: Inferences from TEX86 and Isotope Data," *Geology*, V. 34, pp. 737–40, 2006.

13. Zachos, J., Dickens, G., and Zeebe, R., "An Early Cenozoic Perspective on Greenhouse Warming and Carbon-Cycle Dynamics," *Nature*, V. 45, pp. 279–83, 2008.

14. A review of methane hydrate and its potential role in climate change can be found in Ruppel, C., "Methane Hydrates and Contemporary Climate Change," *Nature Education Knowledge*, V. 3(10), 2011.

15. Kwok, R., & Cunningham, G., "Variability of Sea Ice Thickness and Volume from CryoSat-2," *Philosophical Transactions of the Royal Society A*, V. 373, Article 20140157, 2015.

16. Arctic sea ice has been slow to reform in late 2020, and the sea-ice area reached record monthly lows in October and November. According to the National Snow and Ice Data Center, this strong anomaly is due to the transfer of heat from open water into the atmosphere and a very slow

rate of ice formation on the Russian side of the Arctic Ocean. nsidc.org
/arcticseaicenews, accessed November 2020.

17. Wadhams, P., "The Global Impacts of Rapidly Disappearing Arctic Sea
Ice," *Yale Environment 360*, 2016, accessed September 2020.

18. Lenton, T., et al., "Tipping Elements in the Earth's Climate System," *Proceedings of the National Academy of Sciences*, V. 105, pp. 1786–93, 2008.

19. Grignot, E., et al., "Four Decades of Antarctic Ice Sheet Mass Balance
from 1979 to 2017," *Proceedings of the National Academy of Sciences*, V.116,
pp. 1095–103, 2019.

20. Murton, J., et al., "Preliminary Paleoenvironmental Analysis of Permafrost Deposits at Batagaika Megaslump, Yana Uplands, Northeast
Siberia," *Quaternary Research*, V. 87, pp. 314–30, 2017; Vadakkedath, V.,
Zawadzki, J., and Przeździecki, K., "Multisensory Satellite Observations of
the Expansion of the Batagaika Crater and Succession of Vegetation in Its
Interior from 1991 to 2018," *Environ Earth Sci.*, V. 79, 2020.

21. Turetsky, M., et al., "Permafrost Collapse Is Accelerating Carbon Release,"
Nature, V. 569, pp. 32–4, 2019.

22. Lawrence, D., et al., "Accelerated Arctic Land Warming and Permafrost
Degradation During Rapid Sea Ice Loss," *Geophysical Research Letters*,
V. 35, 2008, L11506.

23. Shakhova, N., et al., "Methane Release on the Arctic East Siberian Shelf,"
Geophysical Research Abstracts, V. 9, 01071, 2007. Methane release from
the Siberian shelf is also the subject of a joint Sweden-Russia study in
2020, described at aces.su.se/research/projects/the-isss-2020-arctic
-ocean-expedition.

24. "A biome is a community of plants and animals that have common
characteristics for the environment they exist in. They can be found
over a range of continents. Biomes are distinct biological communities
that have formed in response to a shared physical climate." "Biome,"
Wikipedia.

25. In Canada, wildfires are most common and most extensive within the
boreal forest. According to Hanes and coauthors, "Results suggest that
large fires have been getting larger over the last 57 years and that the fire
season has been starting approximately one week earlier and ending one
week later.... Area burned, number of large fires, and lightning-caused
fires are increasing in most of western Canada, whereas human-caused
fires are either stable or declining throughout the country. Overall,
Canadian forests appear to have been engaged in a trajectory toward
more active fire regimes over the last half century." Hanes et al., "Fire
Regime Change in Canada over the Last Half Century," *Canadian Journal of
Forest Research*, V. 49, pp. 256–69, 2019.

26. Lenton, et al., "Tipping Elements in the Earth's Climate System," 2008.

27. Curry, R., Dickson, B., and Yashayaev, I., "A Change in the Freshwater Balance of the Atlantic Ocean over the Past Four Decades," *Nature*, V. 426, pp. 826–9, 2003.

28. Lenton, et al., "Tipping Elements in the Earth's Climate System," 2008.

29. "Congressional Research Service: In Focus," September 2020, crsreports .congress.gov. Refer also to figure 10.1.

30. Hanes et al., "Fire Regime Change in Canada," 2019.

31. Veronica Penney, "It's Not Just the West: These Places Are Also on Fire." *New York Times*, September 16, 2020, updated September 23, 2020.

32. Readfearn, G., "Great Barrier Reef's Third Mass Bleaching in Five Years the Most Widespread Yet," *Guardian*, April 6, 2020.

33. IPCC, 2018, "Summary for Policymakers," in *Global Warming of 1.5°C*. An IPCC Special Report on the impacts of global warming of 1.5°C above pre-industrial levels and related global greenhouse gas emission pathways, in the context of strengthening the global response to the threat of climate change, sustainable development, and efforts to eradicate poverty, World Meteorological Organization, Geneva, Switzerland.

34. Turner, A., and Annamalai, H., "Climate Change and the South Asian Summer Monsoon," *Nature Climate Change*, V. 2, pp. 587–95, 2012.

35. Cox, P., et al., "Amazonian Forest Dieback Under Climate-cycle Projections for the 21st Century," *Theoretical and Applied Climatology*, V. 78, pp. 137–56, 2004.

36. Collins, M., et al., "Long-term Climate Change: Projections, Commitments and Irreversibility," in *Climate Change 2013: The Physical Science Basis*, 2013. Contribution of Working Group I to the Fifth Assessment Report of the Intergovernmental Panel on Climate Change, Cambridge University Press.

Chapter 11: What Now?

1. Ruminants (e.g., cows, sheep, goats) produce far more methane than other livestock (e.g., pigs, chickens) because of fermentation within their first two digestive chambers (rumen and reticulum). Most methane is emitted during belching (not farting), but some is produced by manure on the ground or in manure processing facilities. "Ruminant," *Wikipedia*, accessed September 2020.

2. There are many different chlorofluorocarbons (CFCs). A common one is CCl_2F_2, or difluorodichlorocarbon (aka CFC-12). When CFCs were phased out following the Montreal Protocol in 1987, they were replaced by HCFCs that didn't deplete stratospheric ozone as much, but they still acted as greenhouse gases in the lower atmosphere. HCFCs used for refrigeration

and air conditioning are now being replaced with other gases, such as cyclopentane (C_5H_{10}).

3. There are several oxides of nitrogen. Only nitrous oxide (N_2O) is a GHG. The nitrogen oxides NO and NO_2 contribute to fog and acid precipitation but are not GHGs. They are primarily produced during the combustion of fossil fuels.

4. "Most popular cars in America," *Edmunds*, accessed September 2020.

5. Statistics Canada, 150.statcan.gc.ca/t1/tbl1/en/tv.action?pid=201000 0201, accessed September 2020.

6. California New Car Dealers Association, *California Auto Outlook*, V. 16, No. 1, February 2020.

7. Richardson, J., "The Incentives Stimulating Norway's Electric Vehicle Success," *CleanTechnica*, January 28, 2020, accessed September 2020.

8. Ambrose, J., "UK Plans to Bring Forward Ban on Fossil Fuel Vehicles to 2030," *Guardian*, September 21, 2020, accessed November 2020 at theguardian.com; "Phase-out of fossil fuel vehicles," *Wikipedia*, accessed November 2020.

9. U.S. Bureau of Transportation Statistics, bts.gov/statistical-products /surveys/national-household-travel-survey-daily-travel-quick-facts, accessed September 2020.

10. "Solar PV" is the energy from solar photo-voltaic systems, or "solar panels." Levelized cost of energy (LCOE) is the cost of energy production over the lifetime of a system, including capital costs, maintenance and fuel costs, and the amount of energy that can be produced. Of course, the cost of residential solar and the amount of energy that can be produced varies widely from type of installation and region.

11. The Inventory of Carbon and Energy is available at circularecology.com.

12. Karakurt, I., Aydin, G., and Aydiner, K., "Sources and Mitigation of Methane Emissions by Sectors: A Critical Review," *Renewable Energy*, V. 39, pp. 40–8, 2012.

13. In many cases, it is cheaper to burn or "flare" natural gas that is produced along with petroleum than it is to capture and use it. In the process of flaring, methane is converted to carbon dioxide, a less potent GHG than methane (by a factor of 25 to 30 times) that doesn't represent an explosion risk. Flaring is a massive waste of energy, and most petroleum-producing countries are making efforts to curb this practice.

14. From an article at Carbon Brief, carbonbrief.org/explained-fugitive-methane-emissions-from-natural-gas-production (September 2020), showing that fugitive emissions range from just under 1% to over 5%, and average 3% of natural-gas production amounts. The article is based on data from peer-reviewed and government documents.

15. Gossling, S., and Humpe, A., "The Global Scale, Distribution and Growth of Aviation: Implications for Climate Change," *Global Environmental Change*, V. 65, 2020.

16. Quarmby, L., *Watermelon Snow, Science, Art and a Lone Polar Bear*, McGill-Queens University Press, 2020.

17. Liu, Z., et al., "Near-real-time Monitoring of Global CO_2 Emissions Reveals the Effects of the COVID-19 Pandemic," *Nature Communications*, V. 11, 5172, 2020.

18. Friedlingstein, P., et al., "Global Carbon Budget 2020," *Earth System Science Data*, V. 12, pp. 3269–340, 2020.

Index

About the Author

STEVEN EARLE, PhD, has worked in the earth sciences, has developed and taught university earth science courses for almost four decades, and wrote the widely used university textbook *Physical Geology*, now in its second edition. He participates in climate change research and community engagement with climate change solutions, including low-carbon transport initiatives, heating systems, and land stewardship. He and his family live on a sustainable farm on Gabriola Island, BC, Canada.

Additional Resources from New Society Publishers

Living the 1.5 Degree Lifestyle
Why Individual Climate Action Matters More than Ever
LLOYD ALTER

6 × 9" / 176 Pages
US/Can $19.99
PB ISBN 9780865719644

We're All Climate Hypocrites Now
How Embracing Our Limitations Can Unlock the
Power of a Movement
SAMI GROVER

6 × 9" / 192 Pages
US/Can $19.99
PB ISBN 9780865719606

Facing the Climate Emergency
How to Transform Yourself with Climate Truth
MARGARET KLEIN SALAMON

5.5 × 8.5" / 160 Pages
US/Can $14.99
PB ISBN 9780865719415

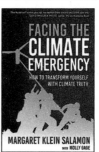

Being the Change
Live Well and Spark a Climate Revolution
PETER KALMUS

6 × 9" / 384 Pages
US/Can $21.99
PB ISBN 9780865718531

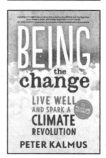

For a full list of titles from New Society Publishers, visit newsociety.com

ABOUT NEW SOCIETY PUBLISHERS

New Society Publishers is an activist, solutions-oriented publisher focused on publishing books to build a more just and sustainable future. Our books offer tips, tools, and insights from leading experts in a wide range of areas.

We're proud to hold to the highest environmental and social standards of any publisher in North America. When you buy New Society books, you are part of the solution!

At New Society Publishers, we care deeply about *what* we publish—but also about *how* we do business.

- All our books are printed on 100% **post-consumer recycled paper**, processed chlorine-free, with low-VOC vegetable-based inks (since 2002). We print all our books in North America (never overseas)

- Our corporate structure is an innovative employee shareholder agreement, so we're one-third employee-owned (since 2015)

- We've created a Statement of Ethics (2021). The intent of this Statement is to act as a framework to guide our actions and facilitate feedback for continuous improvement of our work

- We're carbon-neutral (since 2006)

- We're certified as a B Corporation (since 2016)

- We're Signatories to the UN's Sustainable Development Goals (SDG) Publishers Compact (2020–2030, the Decade of Action)

To download our full catalog, sign up for our quarterly newsletter, and to learn more about New Society Publishers, please visit newsociety.com

---

 ENVIRONMENTAL BENEFITS STATEMENT

New Society Publishers saved the following resources by printing the pages of this book on chlorine free paper made with 100% post-consumer waste.

TREES	WATER	ENERGY	SOLID WASTE	GREENHOUSE GASES
34	**2,700**	**14**	**120**	**14,600**
FULLY GROWN	GALLONS	MILLION BTUs	POUNDS	POUNDS

 Environmental impact estimates were made using the Environmental Paper Network Paper Calculator 4.0. For more information visit www.papercalculator.org

 Certified **B** Corporation

 new society PUBLISHERS www.newsociety.com

 FSC www.fsc.org ·MIX Paper from responsible sources FSC® C016245

 SDG PUBLISHERS COMPACT